The Logica Yearbook
2017

The Logica Yearbook 2017

Edited by

Pavel Arazim

and

Tomáš Lávička

© Individual authors and College Publications 2018
All rights reserved.

ISBN 978-1-84890-281-7

College Publications
Scientific Director: Dov Gabbay
Managing Director: Jane Spurr

www.collegepublications.co.uk

Original cover design by Laraine Welch
Printed by Lightning Source, Milton Keynes, UK

All rights reserved. No part of this publication may be reproduced, stored in a retrieval system or transmitted in any form, or by any means, electronic, mechanical, photocopying, recording or otherwise without prior permission, in writing, from the publisher.

Preface

The book that you are holding in your hands is a further entry in a series of volumes that aspires to make some of the ideas presented at the annual international symposium Logica permanently accessible to both the conference participants and the wider public.

Last year's symposium, which took place at Hejnice Monastery in the Czech Republic from June 19 to June 23, 2017, brought together scholars from many different countries. This volume contains a representative sample of the contributions made at the conference. The Logica symposium is an event with a long tradition, as it was first held in 1987. The symposium is open to researchers of both a mathematical and a philosophical bent.

The informal atmosphere provides a space for a stimulating exchange of ideas among logicians of all generations, including students. As the editors of this volume we are proud that we can contribute to the successful completion of the annual symposium cycle by presenting this collection to you. Last year's Logica was, just as all the previous Logica symposia, organized by the Department of Logic of the Institute of Philosophy of the Czech Academy of Sciences. More than thirty lectures were presented during the conference, including those given by the distinguished invited speakers: Hartry Field, Bob Hale, Shahid Rahman, and Sonja Smets. As is the case every year, the conference was enriched by a social programme that provided room for friendly debates concerning professional topics as well as for starting and
developing personal friendships. The proceedings, which are traditionally published within one year of the conference, unfortunately can only offer a limited record of the topics discussed and cannot hope to even partially convey its atmosphere.

Nevertheless, we believe that the articles in this volume can more than stand on their own. We would especially like to call your attention to the article by Bob Hale, a speaker whom we were particularly happy to welcome at the event. Unfortunately, he passed away in December and so could not complete the article and we therefore had to finalize the editorial work on his article without him. We are grateful to our colleague Ansten Klev for his considerable help in this task, as Bob Hale's article is related to Ansten's own work. Aside from its lack of an abstract, the article is in a great shape and, as saddened as we are by Bob Hale's death, we are honoured to be publishing one of the last articles written by this important figure of contemporary logic and analytical philosophy.

Both the Logica symposium and The Logica Yearbook are the result of a joint effort by many people to whom we would like to express our gratitude. We are, of course, very grateful to the Institute of Philosophy for all the important support that made the event possible. We would also like to express our thanks to the staff of Hejnice Monastery for their hospitality and friendly assistance. Special thanks from the organizers – and also, we are certain, from the guests – must go to the Bernard Family Brewery of Humpolec, which has traditionally sponsored the social programme of the symposium by providing three barrels of its excellent beer. We owe thanks to the Czech Science Foundation, which provided significant support for the meeting and for the publication of this book with funding from the grant
project no. 13-21076S. We would also like to express our gratitude to Olga Bažantová, who is a key member of the organizing crew, and to College
Publications and its managing director, Jane Spurr, for their very consistent and pleasant cooperation during the preparation of this book. Last, but not least, we would like to thank all of the authors for their exemplary collaboration during the editorial process.

Prague, May 2018

Pavel Arazim and Tomáš Lávička

Table of Contents

Causation and Prediction in the Trojan Fly Scenario 1
  Emil Badici

Bases for an Action Logic to Model Negative Modes
of Actions 13
  Ilaria Canavotto

Normality Operators and Classical Collapse 29
  Roberto Ciuni and Massimiliano Carrara

Reconsidering the 'Ingredients' of Explicit Knowledge 47
  Claudia Fernández-Fernández and
  Fernando R. Velázquez-Quesada

On What Counts as a Translation 61
  Alfredo Roque Freire

Extensions and Projections in Deontic Default Logic 77
  André Fuhrmann

What Makes True Universal Statements True? 89
  Bob Hale

Choosing Your Nonmonotonic Logic: A Shoppers Guide 109
  Ulf Hlobil

Towards an Operational View of Purity 125
  Reinhard Kahle and Gabriele Pulcini

A Multi-Succedent Sequent Calculus for Logical
Expressivists 139
  Dan Kaplan

Logical Dialogues in Abstract Argumentation Frameworks 155
  Hanna Karpenko and Olivier Roy

"I Asked You to Mail that Letter, Not to Burn It,"
An Illocutionary Logical Analysis of Directive Acts
and Arguments 169
  John T. Kearns

The Logical Form of Identity Criteria 181
  Ansten Klev

Proof in Mathematics and in Logic 197
  Danielle Macbeth

Laws of Logic – Where Do They All Come From? 209
  Jaroslav Peregrin and Vladimír Svoboda

Ranking Semantics for Doxastic Necessities and Conditionals 223
  Eric Raidl

Frege's *Begriffsschrift* and Logicism 239
  Joan Bertran-San Millán

Reasoning About Fiction 255
  Tom Schoonen and Franz Berto

A Logical Perspective on Social Group Creation 271
  Sonja Smets and Fernando R. Velázquez-Quesada

Causation and Prediction in the Trojan Fly Scenario

EMIL BADICI[1]

Abstract: What makes the Trojan Fly scenario paradoxical is the fact that, contrary to common intuitions about the case, the trajectory of the fly can be proved to be undetermined and unpredictable. Laraudogoitia (2014) argues that the future state of the fly can still be causally determined and predicted when Achilles meets certain requirements pertaining to his disposition and capacity to correctively interfere with the fly's trajectory. This new Trojan Fly scenario is then used to undermine the view that dispositions have causal powers only insofar as they have a categorical basis. This paper aims at showing that the disposition and capacity attributed to Achilles would not suffice to causally determine the behavior of the fly, and that the argument he is supposed to use to predict the state of the fly is circular and thus devoid of any predictive power. Moreover, the new Trojan Fly scenario turns out not to have any bearing on the relation between dispositions and categorical properties.

Keywords: The Trojan Fly paradox, Benardete's paradox of the gods, Zeno, Causation, Prediction

The Trojan Fly paradox is one of the relatively more recent and less known additions to Zeno's family of paradoxes. A familiar example of it goes as follows:[2]

> Achilles travels at 8 mph but the tortoise manages only 1 mph. So Achilles has given it a start. At the point where Achilles catches the tortoise he draws level with a fly which proceeds to fly back and forth between them at 20 mph. After another hour Achilles is 7 miles ahead of the tortoise, but where is the fly? It looks as if it should be possible to calculate its position and determine its direction. But the answer is that it could be anywhere, facing either direction. (Clark, 2002, p. 246)

---

[1] I am indebted to audiences at NM-TX Philosophical Society 2017 and Logica 2017 for useful comments.

[2] The original version of the paradox goes back to A. K. Austin (1971, p. 18).

Emil Badici

It is not hard to construct an argument to the effect that the fly could be anywhere between the two runners and could face either direction. For any such possible state of the fly after one hour, one can trace its path all the way back to the place where Achilles catches the tortoise. What makes the case paradoxical is the conflict between the fully deterministic behavior of the fly described in the scenario and the proof that its trajectory is both undetermined and unpredictable. It should be pointed out that aside from raising a couple of interesting questions concerning its coherence and its relation to other Zeno style paradoxes,[3] this Trojan Fly scenario does not pose a serious challenge to paradox solvers. Since there is no first segment in the fly's path after t_0, the time when Achilles catches the tortoise, the trajectory of the fly remains under-described, which suffices to explain its undetermined and unpredictable behavior. However, this paper is primarily concerned with a modified version of the Trojan Fly due to Laraudogoitia (2014), which combines together the original Trojan Fly paradox and some elements of Benardete's paradox of the gods,[4] a paradox which has often been used to defend the possibility of an unusual type of causation.[5] This combination gives him the opportunity to draw, in the modified case, the conclusion that the behavior of the fly can be both determined and predicted, which is contrary to what can be said about the original scenario. The new Trojan Fly scenario is constructed by introducing a number of additional assumptions

---

[3]There are concerns having to do with the possibility of performing supertasks and concerns with regards to the possibility of instantaneous changes of velocity (recall that the fly is supposed to move at constant speed while instantly reversing direction). I am not going to address these challenges to the coherence of the scenario because they either are not specific to the Trojan Fly paradox or have been satisfactorily addressed by others. See, for instance, Grünbaum (1970) and Salmon (1975). Salmon points out that since the (regressive) Trojan Fly scenario is essentially the reflection in a mirror of Thomson's progressive lamp scenario (see Thomson, 1954), the charges of logical inconsistency should be handled in the same way.

[4]Benardete (1964, pp. 259–60) states the paradox as follows: 'A man decides to walk one mile from A to B. A god waits in readiness to throw up a wall blocking the man's further advance when the man has travelled $1/2$ mile. A second god (unbeknown to the first) waits in readiness to throw up a wall of his own blocking the man's further advance when the man has travelled $1/4$ mile. A third god ...&c. ad infinitum. It is clear that this infinite sequence of mere intentions (assuming the contrary-to-fact conditional that each god would succeed in executing his intentions if given the opportunity) logically entails the consequence that the man will be arrested at point A; he will not be able to pass beyond it, even though not a single wall will in fact be thrown down in his path. The before-effect here will be described by the man as a strange field of force blocking his passage forward'.

Although Laraudogoitia plays down the significance of the connection between his version of the Trojan Fly and the paradox of the gods, the former is undeniably inspired from the latter.

[5]See Hawthorne (2000) and Laraudogoitia (2005).

Causation and Prediction in the Trojan Fly Scenario

pertaining to Achilles' disposition and capacity to act in accordance with a set of conditions C_n (to be specified later). Such dispositions and capacities would allow him (Achilles) to correctively interfere, should the need occur, with the fly's trajectory in order to make sure that it reaches the desired location. Laraudogoitia's view can be summarized in the following two theses:[6]

> (A) Achilles can, by merely having the disposition and capacity to act in accordance with the set of conditions $C_n(n \geq 1)$, causally determine the fly to follow the designated evolution (without actually interfering with its motion).

> (B) If Achilles knows that (A) is true and that he has the disposition and capacity to act in accordance with the set of conditions $C_n(n \geq 1)$, then he is able to predict on the basis of this knowledge that the fly will follow the designated evolution.

The 'designated evolution' refers here to the trajectory Achilles would like the fly to follow. A more precise characterization of this notion will be offered later. The most surprising aspect of thesis (A) is the implication that Achilles can cause the fly to follow the desired path without actually interfering with its movement. If meaningful, this unusual type of causal determination would be philosophically fruitful. For instance, Laraudogoitia points out, the following thesis about the causal powers of dispositions is a corollary of (A):

> (C) Some dispositions have causal powers even though they have no categorical basis (i.e., some events are caused by bare dispositions).

The purpose of this paper is to show that the arguments offered to defend (A) and (B) fail. First, although there is an undeniable connection between Achilles' disposition and capacity and the behavior of the fly, there is no good evidence for thinking of it as a causal connection. Second, the argument that Achilles is supposed to use to predict the state of the fly is devoid of predictive power. For all one knows, Achilles' knowledge that he has the disposition and capacity to act in accordance with the set of conditions C_n might be based on his prior knowledge that the fly follows the designated

---

[6] Although Laraudogoitia does not discuss causation and prediction separately, they are better examined as two different problems.

trajectory. Unless he can find an independent way of knowing that he indeed has the required disposition and capacity, the argument Achilles uses to predict the fly's behavior is circular.

1 The modified Trojan Fly scenario

As the following quote indicates, the modified scenario does not leave out any of the details specified in the original scenario but rather supplements them with additional data:

> [...] I do not question that, given ONLY the data in the supertask of the Trojan fly, the behavior of the fly cannot be predicted (and in this I agree with Gardner and Clark). What I demonstrate is that, given these data, one may have supplementary information available that enables us to predict the evolution of the fly and that this prediction is not therefore impossible. (Laraudogoitia, 2014, p. 774)

Unfortunately, this "supplementary information" cannot be explained without first introducing a few technical terms. Let us focus on the trajectory of the fly between t_0, the time when Achilles, the tortoise and the fly overlap, and t_1 (one hour later), and let us call, with Laraudogoitia, all possible trajectories of the fly which are compatible with the initial scenario 'basic evolutions'. Since the motion of the fly after t_0 is described as a deterministic process, any state of the fly between t_0 and t_1 uniquely determines a basic evolution. In particular, the desired state of the fly at t_1 (i.e., a specific position/direction pair where Achilles would want the fly to be) determines a unique basic evolution, which will be called the 'designated evolution'. Let us now introduce the series S^* of instants corresponding to a basic evolution. If the fly keeps moving at a constant speed, there is an infinite sequence of instants $t_i^* (i \geq 1)$ at which it reaches Achilles:

$$S^*: \quad t_0 < \cdots < t_3^* < t_2^* < t_1^* < t_1.$$

Squeezing infinitely many instants into a finite interval of time should not be a problem since the distance between two contiguous time instances in the series approaches zero as the instances approach the initial time t_0. Notice that every basic evolution determines a unique sequence of instants, and every single instant determines both a unique sequence of instants and a unique basic evolution. As we saw before, a 'designated evolution' is a

Causation and Prediction in the Trojan Fly Scenario

basic evolution that produces the desired state of the fly at time t_1. Suppose at t_0 Achilles desires that at t_1 the fly will be moving towards him and will be positioned at the mid-point between him and the tortoise. This desired state of the fly determines a unique sequence of instants and a unique basic evolution (the designated evolution), which can be backtracked by means of simple classical mechanics methods. Thus, a first thing one needs to assume in the modified scenario is that Achilles is able to calculate the sequence S^* which corresponds to this designated evolution.[7] In addition to this, Achilles must have the disposition to interfere and make "corrections" to the trajectory of the fly in case it does not match the designated evolution. More precisely, he must be disposed to act in accordance with the following set of conditions, where $n \geq 1$:

> C_n: If the fly is at Achilles at an instant $t^\wedge$ such that $t_n^* > t^\wedge > t_{n+1}^*$, then Achilles will keep it with him until the instant t_n^*, the moment in which he will free it to return towards the tortoise (at its velocity, stipulated from the beginning, of 20 mph) without detaining it at any point in the future. (Laraudogoitia, 2014, p. 776)

Finally, since one's disposition to do X does not by itself guarantee that one is going to do X unless one is capable of doing X and, ascending to the epistemic level, since knowledge that one has the disposition to do X does not by itself guarantee knowledge that one is going to do X unless one knows that one is capable of doing X, Laraudogoitia relies on the "implicit assumption" that "Achilles knows that he can act as per the conjunction of all the $C_n(n \geq 1)$, i.e., that he knows he is capable of following the plan detailed in the conjunction of all the $C_n(n \geq 1)$" (ibid., p. 776). It is for this reason that the term 'capacity' has been made part of both (A) and (B). For convenience, the implicit assumption can be split into an epistemic requirement (Kc) and a non-epistemic requirement (c), the latter being a mere consequence of the former (assuming the factivity of knowledge):

(Kc) Achilles knows that he has the capacity to act in accordance with the infinite set of conditions $C_n(n \geq 1)$.

(c) Achilles has the capacity to act in accordance with the infinite set of conditions $C_n(n \geq 1)$.

---

[7] It might be thought that this would be an impossible task. However, Achilles does not have to produce an actual infinite list of time instants in numerical form; all that is needed is the ability to calculate, for each n, the instant t_n.

Anticipating somewhat, it will turn out that the reliance on the implicit assumption is the primary culprit that undermines Laraudogoitia's modified Trojan Fly project. It is now time to examine the way the two theses stated in the introductory section, (A) and (B), are argued for. The former is defended by a reductio argument. Suppose that Achilles has the disposition and capacity to act in accordance with the set of conditions $C_n (n \geq 1)$. Let us now assume, for reductio, that for some value n the fly arrives at Achilles at an instant $t^\wedge$, such that $t_n^* > t^\wedge > t_{n+1}^*$ (i.e., we assume that the path of the fly does not coincide with the designated evolution). If so, the fly must also have arrived at Achilles at an instant $t^{\wedge\wedge}$, such that $t_{n+1}^* > t^{\wedge\wedge} > t_{n+2}^*$. This, however, could not have happened because at $t^{\wedge\wedge}$ Achilles would have kept the fly with him and would have released it at t_{n+1}^*, in which case he (the fly) would not have arrived at Achilles at $t^\wedge$, contrary to the initial assumption. Notice that in order to infer that at $t^{\wedge\wedge}$ Achilles would have kept the fly with him and would have released it at t_{n+1}^*, one needs to rely on (c). Once the reductio is completed, one is able to infer that "the evolution of the fly will necessarily be the designated evolution" (ibid., p. 776).[8] Laraudogoitia makes it clear that the relation his proof brings out is a causal relation: "it was the disposition of Achilles to act in accordance with the conjunction of all the C_n that caused the fly to end up after an hour at the mid-point between him and the tortoise and moving towards him" (ibid., p. 779). If the argument offered above is enough to justify the causation thesis, then all that Achilles needs to do to causally determine the fly to follow the designated evolution is calculate the corresponding sequence of instants S^* and be disposed and capable to act in accordance with the set of conditions C_n. Moreover, since there is no time after t_0 when the fly's trajectory is in need of correction, Achilles does not have to interfere with it in any way. His bare dispositions are enough to produce, through some "strange causal power" (ibid., p. 777), the desired effect. Consider now thesis (B). Although one's ability to causally determine an event does not by itself guarantee one's ability to predict its occurrence, it can guarantee it together with some epistemic requirements. First, Achilles must know that (A) is the case. Second, he must know that he has the disposition to act in accordance with

---

[8]One could object that Laraudogoitia changes the problem when he allows Achilles to keep the fly to himself for a certain interval of time. While Achilles is holding it, the fly's velocity equals the velocity of Achilles himself, which is contrary to the original scenario in which the fly's velocity is constant over the entire process. However, this objection would be misguided, because in the modified scenario Achilles does not in fact make use of his power. The speed of the fly remains constant.

the set of conditions C_n. From these, together with (Kc), he is able to infer that the fly will follow the designated trajectory.[9] As far as the last thesis, (C), is concerned, it can be easily inferred as a corollary of (A). If (A) is true, then the behavior of the fly is caused either by a categorical property or by a bare disposition. If the cause is a categorical property, there must be some microstructural property of Achilles that causes the fly to move in the right way in virtue of some laws of nature. However, since Achilles never interferes with the fly, this cannot be the case. It follows that the behavior of the fly must be caused by a bare disposition.

2   Causation and prediction refuted

If true, the claim that Achilles can causally determine and predict the trajectory of the fly would be highly significant. Unfortunately, neither of the two theses is justified. Let us first examine (A). Recall that Laraudogoitia uses the reductio proof to show that if Achilles has the disposition and capacity to act in accordance with the set of conditions C_n, then the fly follows the designated evolution. In other words, it shows that, assuming that (d) and (f) stand for the first two sentences below, the counterfactual conditional (cc) is true:

(d) Achilles has the capacity to act in accordance with the infinite set of conditions $C_n (n \geq 1)$.

(f) The fly follows the designated trajectory.

(cc) If (d) and (c) were true, then (f) would have been true.

At first glance, this counterfactual dependence might seem to be enough to guarantee causal dependence. Although Laraudogoitia does not offer any details on either the nature of the two relata or on the specific type of causal theory he favors, it is not hard to see how the argument above can be adapted to conform to a counterfactual theory of causation. For Lewis (1973), two distinct events, e_1 and e_2, are causally dependent if and only if the following

---

[9]Laraudogoitia assumes here that "someone's disposition to act as per plan X implies that, if he is capable of acting in accordance with X, he will do so" (ibid., p. 776). It can be objected that even if one is disposed and capable to act according to X, one might fail to act if there are overriding reasons not to. One could have the disposition and the ability to punch one's rival in his face but still refrain from doing so for moral, legal or other overriding reasons. Thus, some qualifications are necessary in order for the argument to go through, but there is no reason to think that this is a crucial impediment.

two counterfactuals are true: i) if e_1 had occurred e_2 would have occurred and ii) if e_1 had not occurred e_2 would not have occurred. Since the new Trojan Fly scenario is presumably considered to be a merely counterfactual possibility, the first counterfactual is trivially true, while the second is the conditional (cc) mentioned above. Thus, there seems to be enough evidence for a causal dependence relation.

However, counterfactual dependence does not always mean causal dependence, as Lewis himself explicitly claimed. For instance, he argues, since "John's saying 'Hello' loudly" implies "John's saying 'Hello'", there is a clear counterfactual dependence between the events described by the two expressions. If the former had occurred, then the latter would have occurred as well. However, this counterfactual dependence is not an instance of causal dependence because "[w]e may take it as a general principle that when one event implies another, then they are not distinct and their counterfactual dependence is not causal" (Lewis, 1986, p. 256).[10] The argument offered by Laraudogoitia fails to prove causal dependence for a similar reason. Recall that since (d) is not enough to prove (f), the reductio proof needs to rely on (c) as well. The problem is that this implicit assumption is not as innocuous as it might seem. Capacities or abilities are intricate properties that lend themselves to several possible interpretations. I will argue that even if (c) is given a coherent reading, it fails to support the intended causal thesis.

Consider, as an analogy, the example of Smith, an expert burglar who knows how to unlock the door of my safe without keys, and Jones, a novice who is yet unable to do it. While the truth of (s) is independent on whether the safe is actually locked, things are less clear when it comes to (j), which allows at least two significantly different readings.

(s) Smith has the capacity to open my safe door without keys.

(j) Jones has the capacity to open my safe door without keys.

The broad reading of (j) interprets it as referring to what Jones could do in a broad range of possible circumstances in which the safe door could be locked or unlocked. The narrow reading interprets it as referring to what he is able to do in the actual circumstances. According to the broad reading, (j) is false regardless of whether the door is actually unlocked. On the

---

[10] For Lewis, 'different' and 'distinct' are not synonymous. Two different events can fail to be distinct if their spatio-temporal regions include one another.

other hand, according to the narrow reading, (j) is true if the door is actually unlocked and false otherwise. When interpreted in the narrow sense, the door's being unlocked is a precondition for Jones' having the capacity to open it without keys. While it is true that if Jones had the desire and (narrow) capacity to open my safe door, the door would have been unlocked, it would be wrong to use this counterfactual as evidence of a causal connection. This case however is different from the Trojan Fly case, because in the latter Achilles is supposed to act in accordance with a set of conditional rules (which cannot be jointly satisfied unless their antecedents are false). A more analogous case would have Jones act in accordance with the following rule:

(R) If the safe door is locked, unlock it.

Jones clearly does not have the (broad) capacity to act in accordance with (R), but he might have the (narrow) capacity to do it (if the door happens to be unlocked). The truth of the counterfactual (R_1) is insufficient to prove a causal relation in which the door's being unlocked is the second *relatum*.

(R_1) Had Jones had the desire and capacity to act in accordance with (R), the door would have been unlocked.

For the first relatum, two possible candidates are Jones' desire to act in accordance with (R) and Jones' desire and capacity to act in accordance with (R). Using (R_1) to prove that Jones' desire to act in accordance with (R) causes the door's being unlocked would be question-begging, because the assumption that Jones has the capacity to act in accordance with (R) already presupposes the door's being unlocked. On the other hand, since the conjunction of Jones' desire and capacity to act in accordance with (R) implies the door's being unlocked, it cannot cause it because Lewis' logical independence and distinctness requirements are not met.

Switching back to the Trojan Fly scenario, Achilles lacks a broad capacity of acting in accordance with the set of conditions C_n whether or not the fly follows the designated trajectory. This is a consequence of the reductio proof. On the other hand, he can have the narrow capacity of acting in accordance with the set of conditions C_n under those possible circumstances in which the fly follows the designated trajectory (and thus it leaves nothing for Achilles to do). Just like in the example above, using the counterfactual (cc) as evidence for a causal relation would be unjustified. If the first relatum is Achilles' disposition to act in accordance with C_n, reliance on (cc) would be question-begging because (c) already presupposes the truth of (f) even when (d) is false. Achilles' disposition to act in accordance with the C_n would be superfluous. On the other hand, the complex state of affairs described by the conjunction of (d) and (c) cannot be the cause of the state of affairs described by (f), because Lewis' logical independence and distinctness requirements are not met.

Consider now thesis (B). Rejecting (A) is not sufficient to undermine the argument for (B) because even if Achilles' disposition and capacity to act in accordance with the C_n has no causal priority over the fly's following the designated trajectory, it might still have epistemic priority. It is true that from his knowledge that he has the disposition and capacity to act in accordance with the conjunction of all the $C_n (n \geq 1)$ Achilles can derive, based on logic alone, the future evolution of the fly. Still, this does not suffice to secure Achilles' ability to make the prediction merely on the basis of what is specified in the new Trojan Fly scenario. If he does not have an independent way of knowing that he has the disposition and capacity (which does not presuppose that he already knows that the fly follows the designated evolution), then he cannot make a genuine prediction. I take it that an analysis of Achilles' ability to predict future event Y on the basis of current knowledge X would have to include at least the following requirements: i) Achilles knows that X, ii) Achilles is able to derive Y from X, and iii) Achilles' knowledge that X is not based on prior knowledge of Y. The third requirement is necessary to ensure that Achilles won't be deriving something that he already knows (in which case it won't be a genuine case of prediction). An example can help make the need for this requirement more vivid. Suppose that Harry is to take a translation test tomorrow, and that the test administrators could select either a text in German (which he has studied in the past) or a text in Chinese (which he has never studied before). Suppose that Harry learns through bribery that the text will be in German and thus, with relief, comes to know that he is capable of passing the test. From the fact that he is capable of passing the test, he is certainly able to derive the fact that the language of the text will be German, but this does not mean that he can predict the language on the basis of his knowledge that he can pass the test, since requirement iii) is violated. The same requirement is violated in the new Trojan Fly scenario. Unless Achilles can find an independent way of knowing that he indeed has the required disposition and capacity, the argument he uses to predict the fly's behavior is circular. If there is another way of knowing it, this must go beyond what is given in the modified Trojan Fly scenario.

References

Benardete, J. (1964). *Infinity: an Essay in Metaphysics*. Oxford: Clarendon Press.
Clark, M. (2002). *Paradoxes from A to Z*. London and New York: Routledge.
Grünbaum, A. (1970). Modern science and Zeno's paradoxes of motion. In W. Salmon (Ed.), *Zeno's Paradoxes* (pp. 200–250). Indianapolis: Bobbs-Merrill.
Hawthorne, J. (2000). Before-effect and Zeno causality. *Nous, 34,* 622–633.
Laraudogoitia, J. P. (2005). Action without interaction. *Analysis, 65,* 140–143.
Laraudogoitia, J. P. (2014). Dispositions and the trojan fly. *Nous, 48,* 773–780.
Lewis, D. (1973). Causation. *The Journal of Philosophy, 70,* 556–567.
Lewis, D. (1986). Events. In D. Lewis (Ed.), *Philosophical Papers* (Vol. 2, pp. 241–269). Oxford University Press.
Salmon, W. (1975). *Space, Time, and Motion*. California: Dickenson Publishing Co. Encino and Belmont.
Thomson, J. F. (1954). Tasks and super-tasks. *Analysis, 15,* 1–13.

Emil Badici
Texas A&M University – Kingsville
USA
E-mail: emil.badici@tamuk.edu

Bases for an Action Logic to Model Negative Modes of Actions

ILARIA CANAVOTTO[1]

Abstract: Currently available systems of action deontic logic are not designed to model procedures to assess the conduct of an agent which take into account the intentions of the agent and the circumstances in which she is acting. Yet, procedures of this kind are essential to determine what counts as culpable not doing. In light of this, we design an action logic, **AL**, in which it is possible to distinguish actions that are objectively possible for an agent, *viz.* there are no objective impediments for the agent to do them, and actions that, besides being objectively possible, are compatible with the setting or intentions of the agent.

Keywords: Action deontic logic, Intentional actions, Omission

1 Introduction

In most juridical systems, but also in everyday practice, the conduct of an agent is typically assessed by answering two fundamental questions:

1. Does the act of the agent comply with the law?

2. Can the agent be held responsible for her act?

By giving a negative answer to the first question, the act of the agent is qualified as a wrong act (or *actus reus*). By giving a positive answer to the second question, the agent is ascribed culpability or blameworthiness for her act, where culpability is usually attributed on the basis of two main criteria:

2.1. a *subjective criterion*, i.e. whether or not the agent really intended to do what she did.

---
[1] I would like to thank Alexandru Baltag, Franz Berto, Alessandro Giordani, Sonja Smets, and the audience of the Logica 2017 conference for comments and helpful feedback.

2.2. a *contextual criterion*, i.e. whether or not the circumstances provided the agent with excuses for her behaviour.[2]

Interestingly, most of the currently available systems of action deontic logic (see e.g. Broersen, 2004; Castro & Maibaum, 2009; Meyer, 1988; Segerberg, 1982; Trypuz & Kulicki, 2015) aim at modelling the deontic status of a given system of actions, independently of the intentions of the agent and of the concrete circumstances in which she is acting, thus disregarding subjective and contextual criteria altogether. As a consequence, from the point of view of these systems, the conduct of the agent can only be assessed from a highly abstract perspective. Taking these criteria into account would allow us not only to avoid some well-known paradoxical consequences of these systems (as recently discussed for example in Canavotto & Giordani, n.d.; Giordani & Canavotto, 2016) but also to improve the characterization of central deontic notions essentially involving the responsibility of the agent, like the notion of omission. This notion is of special importance to deontic reasoning because it lies at the heart of the study of what should count as *culpable not doing*.

The aim of this paper is to lay down the basis for the development of a system of action deontic logic in which the above mentioned improvement can be carried out. We do this by presenting a basic action logic, **AL**, in which a fundamental distinction between actions that are executable by an agent given the circumstances and actions that are executable by an agent given her setting or intentions is modelled. This logic is a variant of the action logic on which the deontic system **ADL** proposed by Canavotto and Giordani in (Canavotto & Giordani, n.d.) is based. With respect to this system and to other action logics inspired by Propositional Dynamic Logic (PDL) and used in deontic logic (esp. Broersen, 2004; Castro & Maibaum, 2009; Meyer, 1988), one of the main conceptual novelties of **AL** is that, among the set of individual actions available to an agent at a given world, we specifically consider those that are compatible with her setting or intentions at that world.

The paper is structured as follows. In the next section, we introduce the basic intuitions on which our system is based and the main distinctions it

---

[2]The precise way in which these criteria are specified and applied to concrete cases is a delicate object of discussion in legal theory (for an example see Botterell, 2009). For our aims, however, an intuitive understanding of the general distinction between subjective and contextual criteria will suffice. We refer the interested reader to (Fletcher, 1998, ch. 5) and to the Model Penal Code to see how these distinctions are typically used in legal texts.

Bases for an Action Logic

aims at capturing. In section 3, we present the syntax and semantic of the action logic **AL**, and we prove a characterization theorem for it. Finally, section 4 concludes by pointing to some key developments.

2 Framing the system

Our proposal is based on the idea that, in order to capture the notion of culpability, more distinctions than those drawn by standard action deontic logics need to be made. In order to introduce the conceptual elements characterizing our framework, let us consider a simple real-life situation.

Suppose that it is 7:30 in the morning and a man, John, while getting ready to go to the office, suddenly remembers that he and his wife promised to their lawyer, Mr Brown, that they would have brought some documents to his law firm by 8:00. They assured him that, if they could not make it in the end, they would have called him by 7:45 to postpone the meeting. In this situation, there are then three main things that John can do, namely: α_1, going straight to the office; α_2, calling the lawyer and postpone the meeting; α_3, going to the law firm with the documents. These three actions are *action types* that can be instantiated in several ways. In particular, we can suppose that John has two principal ways to go the law firm, *viz.* either by car or by bike. If he took the car, John would be at the law firm at 7:45; if he took the bike he would be there at 7:55.

As shown in figure 1, we can think of *instances of actions*, intended as action types, as transitions between possible worlds. By performing different actions, different *states* can be realized at different worlds. For example, in the figure, the state that John is at the law firm at 7:45 is realized at w_3 and the state that he is there at 7:55 is realized at w_4.[3]

Given a certain possible world, we can distinguish two key senses in which an action can be performed by an agent at that world. To see this, imagine, on the one hand, that John's phone is broken. In this situation, the man cannot perform the action α_2 of calling Mr Brown and postpone the meeting, because there is a preventive factor, or *impediment*, in the actual world that prevents him from doing so, namely his phone's being broken.

---

[3] Although in the literature about transition systems the terms "world" and "state" are used interchangeably, here we use them to express different concepts. Specifically, we use "state" to denote a state of affairs and "world" to denote a complete possible world, where a state of affairs can obtain at different possible worlds. Later on, we will model states as sets of worlds, i.e. the sets of worlds where the states obtain. In the semantics for **AL**, states will thus correspond to so-called UCLA propositions.

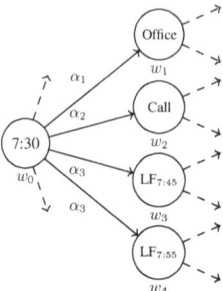

Figure 1: Action Types and Tokens

Since, *given the circumstances*, α_2 is not executable by John, the man has an *excuse* for not calling the lawyer in case he does not go to the law firm. In general, we will say that an action type β is *objectively possible* for an agent just in case there are no impediments for the agent to perform β, where, roughly, an impediment for an agent to perform β is any "non-mental" act or event that prevents the agent from currently performing β.[4]

On the other hand, imagine that there is nothing in the actual world preventing John from going to the law firm, but that he has an important video conference that he intends not to miss. Let us consider two possible scenarios.

- *Scenario 1.* The conference is at 9:30. If John went to the law firm, he would be in time for the conference, no matter whether he goes by car or by bike.

- *Scenario 2.* The conference is at 8:30. If John went to the law firm, he would miss the conference, no matter whether he goes by car or by bike.

Now, unless he changes his mind, John cannot perform any action that makes him miss the conference meeting, in the sense that executing

---

[4]Some examples: a traffic jam might be an impediment for the agent to catch a train; a disease might be an impediment for her to go to work; the lack of necessary means is an impediment for her to achieve her planned end. On the other hand: believing that it is wrong to assume drugs is not an impediment to assume drugs; the fact that a kid is convinced by her mother not to eat chocolate is not an impediment for her to eat chocolate; knowing that crimes are punished is not an impediment to commit a crime.

an action of this sort is incompatible with his intention: in this case, fulfilling his intention excludes the possibility of performing any such action. Hence, *given his setting or intention*, it is possible for John to execute the action α_3 of going to the law firm in scenario 1, but not in scenario 2, even though in neither scenario there are impediments for him to perform this action. The two cases are illustrated in figure 2, where arrows represent instances of actions compatible with John's intention.

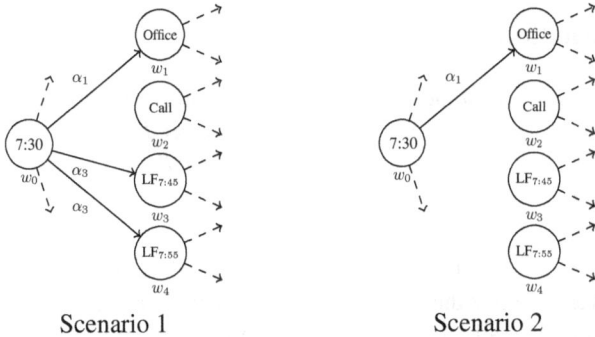

Scenario 1   Scenario 2

Figure 2: Specified scenarios.

Note that, in neither scenario, the action of calling the lawyer is executable by John given his intentions. This depends on the fact that, since his phone is broken, this action is not executable by John at all. Furthermore, it is important to observe that the actions compatible with John's intention might not include the action that John will end up doing. In fact, something could prevent John from carrying out any of these actions (for instance, John might be involved in a car accident).

In general, we will say that an action type, β, is executable by the agent *given her intentions*, just in case, besides being executable by the agent given the circumstances, it can also be performed by the agent in a way that is compatible with her setting.[5] In turn, a way to perform β is compatible with the agent's setting just in case it does not lead to a world where the agent has realised a state contrasting with her intentions. So, actions that are *not* executable by the agent given her setting are either objectively impossible for the agent (as action α_2 in our specified case study) or objec-

---

[5]We use the term "setting" to indicate the entirety of intentions that an agent has in a given situation. Hence, if an agent forms an intention to do an action α in a certain situation, then the setting of that agent in that situation will include an intention to do α.

tively possible for the agent but excluded in virtue of her setting (as action α_3 in scenario 2 of our specified case study). It is worth noting that, since it is possible that the setting of the agent has an impact not only on the current situation but also on all future ones, it can turn out that an action is *permanently* unexecutable by an agent given her setting.

We have now all conceptual elements we need to introduce the logic **AL**.[6]

3 The action logic AL

The language $\mathcal{L}_{\mathbf{AL}}$ of the action logic **AL** includes two sets of expressions, namely a set $Tm(\mathcal{L}_{\mathbf{AL}})$ of action terms and a set $Fm(\mathcal{L}_{\mathbf{AL}})$ of formulas. Let $\mathcal{A}$ be a fixed set of action type variables. $Tm(\mathcal{L}_{\mathbf{AL}})$ is then built according to the following grammar:

$$\alpha ::= a_i \mid 1 \mid \alpha \sqcup \alpha \mid \alpha \sqcap \alpha$$

where $a_i \in \mathcal{A}$. Intuitively, 1 is the action type instantiated by any action whatsoever; $\alpha \sqcup \beta$ is the action type instantiated by any action instantiating either α or β; finally, $\alpha \sqcap \beta$ is the action type instantiated by any action instantiating α and β in parallel. We assume that an individual action can be a token of different types. Hence, saying that an action is a token of a_i does not exclude the possibility that it is also a token of a different type a_j.

Turning to the set of formulas of $\mathcal{L}_{\mathbf{AL}}$, let us fix a countable set $\mathcal{P}$ of propositional variables. Then, $Fm(\mathcal{L}_{\mathbf{AL}})$ is built according to the following grammar:

$$\phi ::= p_i \mid \neg\phi \mid \phi \wedge \phi \mid [F]\phi \mid [1]\phi \mid done(\alpha) \mid imp(\alpha)$$

where $p_i \in \mathcal{P}$ and $\alpha \in Tm(\mathcal{L}_{\mathbf{AL}})$. The other connectives and the dual modalities $\langle F \rangle$ and $\langle 1 \rangle$ are defined as usual. The intended interpretation of the modal formulas is as follows. $[F]\phi$ says that ϕ holds in all worlds that are accessible in the future from the current world, while $[1]\phi$ says that ϕ holds in all the worlds that the agent can reach by acting in a way compatible with her setting or intentions. Finally, $done$ and imp are modal properties of actions such that $done(\alpha)$ is true at all the worlds where the agent has just performed α and $imp(\alpha)$ is true at all the worlds where there are impediments for the agent to do α. As we will see in a moment, the modality $[1]\phi$ and the modal property $done(\alpha)$ can be used to define the familiar action modality $[\alpha]\phi$.

---

[6] As mentioned in the introduction, the logic **AL** is a variant of the action logic used in (Canavotto & Giordani, n.d.). The main differences are the intended interpretation of the action modalities $[\alpha]$ and the presence of the modal property of actions imp.

3.1 Semantics

The semantics for $\mathcal{L}_{\mathbf{AL}}$ is based on the idea that actions are action types that can be performed in different ways and, hence, be instantiated by different individual actions. We standardly model action types as binary relations over possible worlds. Thus, an individual action will correspond to a transition linking two worlds. As in (Canavotto & Giordani, n.d.), we conceive of actions as achievements and accomplishments in Vendler's classification (see Vendler, 1957), or as acts and achievements in von Wright's conceptualization (see von Wright, 1963, 1981). On this conception, there is a perfect correspondence between worlds to which an action leads and worlds where *the state of affairs that the action has been done* is realized. Later on, we will rely on this correspondence to define actions in terms of states.

Definition 1 (Frame) *A frame for $\mathcal{L}_{\mathbf{AL}}$ is a tuple $F = \langle W, R_F, R_1, D, I \rangle$, where*

- *$W \neq \emptyset$ is a set of possible worlds*
- *$R_F : W \to \wp(W)$*
- *$R_1 : W \to \wp(W)$*
  $D : Tm(\mathcal{L}_{\mathbf{AL}}) \to \wp(W)$
- *$I : Tm(\mathcal{L}_{\mathbf{AL}}) \to \wp(W)$*

Let us consider each element in turn. Firstly, R_F is a function that determines, for each world w, which worlds are accessible in the future from w. This function is characterized by the following conditions.

Conditions on R_F: for all $w, v, u \in W$
(a) $w \in R_F(w)$
(b) $v \in R_F(w) \Rightarrow R_F(v) \subseteq R_F(w)$

Intuitively, $R_F(w)$ is thus the cone containing the stages that are accessible from w.

Secondly, R_1 is the function that, for each possible world w, returns the outcomes of all the transitions starting at w which are compatible with the agent's intentions. Hence, intuitively, $R_1(w)$ is the set of worlds the agent can access by acting in a way that is compatible with her setting or intentions. We require that R_1 satisfies the following condition.

Conditions on R_1: for all $w \in W$
(a) $R_1(w) \subseteq R_F(w)$

Hence, worlds that an agent can access by executing an individual action compatible with her setting are worlds that can be accessed in the future.

Observe that the condition on R_1 leaves the possibility open that, for some world w, $R_1(w) = \emptyset$. This means that there might be worlds where the agent's setting excludes the possibility of acting. This happens, for instance, when the agent has conflicting intentions.

Thirdly, D is the function that determines the set of worlds at which a given action has just been performed. Accordingly, for any action α, $D(\alpha)$ is the set of worlds where α has just been done by the agent. This function must satisfy three conditions, which provide a straightforward connection between the algebra of action types and the algebra of states corresponding to the instantiations of the action types:

Conditions on D:
(a) $D(1) = W$
(b) $D(\alpha \sqcup \beta) = D(\alpha) \cup D(\beta)$
(c) $D(\alpha \sqcap \beta) = D(\alpha) \cap D(\beta)$

Hence: (a) some action is instantiated at any stage, (b) instantiating the disjunction of two actions is the same as instantiating either the first or the second action, and (c) instantiating the conjunction of two actions is the same as instantiating both the first and the second action.

Finally, I is the function that determines the set of worlds where there are some impediments for the agent to do α. Hence, $I(\alpha)$ is the set of worlds where there are factors preventing the execution of α. This function is characterized by the following conditions.

Conditions on I: for each $w \in W$
(a) $I(1) = \emptyset$
(b) $I(\alpha \sqcup \beta) = I(\alpha) \cap I(\beta)$
(c) $I(\alpha \sqcap \beta) = I(\alpha) \cup I(\beta)$
(d) $w \in I(\alpha) \Rightarrow R_1(w) \cap D(\alpha) = \emptyset$

According to the conditions on I, (a) there are never impediments to perform all types of action, (b) there are impediments to perform $\alpha \sqcup \beta$ just in case there are impediments to perform both α and β, (c) there are impediments to perform $\alpha \sqcap \beta$ just in case there are impediments to perform either α or β. Finally, in line with the characterization provided in section 2, (d) if, in a certain situation, α is not executable by the agent given the circumstances, then, in that situation, α is not executable by the agent given her setting either (this reading of the consequent of (d) will become clear in a moment).

At this point, we are able to introduce the definition of the key functions of our action deontic logic.

Bases for an Action Logic

Definition 2 (α-transitions) *For every $\alpha \in Tm(\mathcal{L}_{\mathbf{AL}})$, R_α is a map $R_\alpha : W \to \wp(W)$ such that, for each $w \in W$, $R_\alpha(w) = R_1(w) \cap D(\alpha)$.*

R_α is a function that, for each possible world w, returns the outcomes of the transitions starting at w associated with α and compatible with the agent's intentions. More specifically, definition 2 tells us that the transitions associated with α and compatible with the setting of the agent are those transitions compatible with the setting of the agent which end in a world in which the agent has just done α. As we mentioned at the beginning of this section, the basic idea is that there is a perfect correspondence between worlds to which an action leads and worlds where *the state of affairs that the action has just been done* is realized. Crucially, note that the characterization of R_1, $done$ and imp allows for situations where $R_\alpha(w) = \varnothing$, even though $w \notin I(\alpha)$. This is essential to account for the fact that, in certain situations, no way of doing α is compatible with the setting of that agent, even though there is no factor preventing the agent from doing α (see the action of going to the law firm in scenario 2 of our case study).

Definition 3 (Model) *A model for $\mathcal{L}_{\mathbf{AL}}$ is a tuple $M = \langle F, V \rangle$, where*

- *F is a frame for $\mathcal{L}_{\mathbf{AL}}$*
- *$V : \mathcal{P} \to \wp(W)$ is a standard valuation mapping propositional variables into the sets of possible worlds at which they are true.*

Definition 4 (Truth in a model) *The notion of truth of a formula at a world in model for $\mathcal{L}_{\mathbf{AL}}$ is recursively defined as follows.*

$M, w \models p \Leftrightarrow w \in V(p)$
$M, w \models \neg \phi \Leftrightarrow M, w \not\models \phi$
$M, w \models \phi \wedge \psi \Leftrightarrow M, w \models \phi$ and $M, w \models \psi$
$M, w \models [F]\phi \Leftrightarrow \forall v \in W(v \in R_F(w) \Rightarrow M, v \models \phi)$
$M, w \models [1]\phi \Leftrightarrow \forall v \in W(v \in R_1(w) \Rightarrow M, v \models \phi)$
$M, w \models done(\alpha) \Leftrightarrow w \in D(\alpha)$
$M, w \models imp(\alpha) \Leftrightarrow w \in I(\alpha)$

The main operators of our action logic can now be explicitly defined.

Definition 5 (Action modalities) $[\alpha]\phi := [1](done(\alpha) \to \phi)$
The dual modality $\langle \alpha \rangle \phi$ is defined accordingly.

Intuitively, $[\alpha]\phi$ says that ϕ is true at all the worlds that the agent can reach by executing any individual action of type α compatible with her setting. Hence, letting $\top$ be any tautology, the intended meaning of the formula

$\langle \alpha \rangle \top$ is that the agent can perform α in a way that is compatible with her setting. The basic distinction between actions that are objectively possible for an agent and actions that are executable by an agent given her setting can then be expressed in our language by the distinction between $\neg imp(\alpha)$ and $\langle \alpha \rangle \top$. What is more, letting $\bot$ be any contradiction, actions that are permanently unexecutable by an agent given her setting can be characterized as those satisfying the formula $[F][\alpha]\bot$.

3.2 Axiomatization

The axiom system **AL** consists of the following four groups of axioms and rules.

Group 1: axioms and rules for $[F]$
A1.1 $[F](\phi \to \psi) \to ([F]\phi \to [F]\psi)$
A1.2 $[F]\phi \to \phi$
A1.3 $[F]\phi \to [F][F]\phi$
R1.1 $\phi \, / \, [F]\phi$

Group 2: axioms and rules for $[1]$
A2.1 $[1](\phi \to \psi) \to ([1]\phi \to [1]\psi)$
A2.2 $[F]\phi \to [1]\phi$

Group 3: axioms for $done$
A3.1 $done(1)$
A3.2 $done(\alpha \sqcup \beta) \leftrightarrow done(\alpha) \vee done(\beta)$
A3.3 $done(\alpha \sqcap \beta) \leftrightarrow done(\alpha) \wedge done(\beta)$

Group 4: axioms for imp
A4.1 $\neg imp(1)$
A4.2 $imp(\alpha \sqcup \beta) \leftrightarrow imp(\alpha) \wedge imp(\beta)$
A4.3 $imp(\alpha \sqcap \beta) \leftrightarrow imp(\alpha) \vee imp(\beta)$
A4.4 $imp(\alpha) \to [1]\neg done(\alpha)$

Corollary 1 *The following propositions are derivable in* **AL**.
C1.1 $[\alpha \sqcup \beta]\phi \leftrightarrow [\alpha]\phi \wedge [\beta]\phi$
C1.2 $[\alpha]\phi \vee [\beta]\phi \rightarrow [\alpha \sqcap \beta]\phi$
C1.3 $\langle \alpha \rangle \top \leftrightarrow \langle 1 \rangle \, done(\alpha)$

Proof. Immediate from axioms of groups 3 and definition of $[\alpha]\phi$. □

According to **C1.1** and **C1.2**, our action modalities behave as standard action modalities in Dynamic Deontic Logic (see Meyer, 1988). In addition, due to **C1.3**, the executability of an action by an agent given her setting, expressed by $\langle \alpha \rangle \top$, is to be distinguished from the objective concrete possibility the agent has to perform that action, expressed by $\neg imp(\alpha)$. In fact, while $\langle \alpha \rangle \top \rightarrow \neg imp(\alpha)$ is derivable, it is possible that $\neg imp(\alpha)$ holds, even if α is not executable by the agent given her setting.

Theorem 1 *The system* **AL** *is sound and complete with respect to the class of all models for* **AL**.

Proof. The proof of soundness is straightforward. The proof of completeness follows from the proof of the fact that any **AL**-consistent set of formulas is satisfiable in a model for $\mathcal{L}_{\mathbf{AL}}$, which in turn derives from the definition of a canonical model. The construction is as follows.

Definition 6 (Canonical model) *The canonical model for $\mathcal{L}_{\mathbf{AL}}$ is the tuple*
$$M^c = \langle W, R_F, R_1, D, I, V \rangle$$
such that

- W is the set of all maximal **AL**-consistent sets of formulas of $\mathcal{L}_{\mathbf{AL}}$
- $R_F : W \rightarrow \wp(W)$ is such that $v \in R_F(w) \Leftrightarrow w/[F] \subseteq v$, where $w/[F] = \{\phi \mid [F]\phi \in w\}$
- $R_1 : W \rightarrow \wp(W)$ is such that $v \in R_1(w) \Leftrightarrow w/[1] \subseteq v$, where $w/[1] = \{\phi \mid [1]\phi \in w\}$
- $D : Tm(\mathcal{L}_{\mathbf{AL}}) \rightarrow \wp(W)$ is such that $D(\alpha) = \{w \mid done(\alpha) \in w\}$
- $I : Tm(\mathcal{L}_{\mathbf{AL}}) \rightarrow \wp(W)$ is such that $I(\alpha) = \{w \mid imp(\alpha) \in w\}$
- $V : \mathcal{P} \rightarrow \wp(W)$ is such that $V(p) = \{w \mid p \in w\}$

The proof that the canonical model is indeed a model for $\mathcal{L}_{\mathbf{AL}}$ that verifies, at every world, all formulas belonging to that world is routine and is left to the reader.

□

4 Applications and developments

The action logic **AL** is a system in which basic *contextual criteria* required to assess whether or not an agent is responsible or culpable for not having done an action are accounted for. To be sure, suppose that we restrict our attention to actions the agent is able to perform, as it is often done in action deontic logics. Then, the formula $imp(\alpha)$ expresses the fact that the agent has an excuse for not doing α: despite her ability to do α, the actual circumstances keep her from doing it. Besides this, we have seen that **AL** also allows us to analyse the conduct of the agent by considering what she can do given her setting or intentions. Is this sufficient to account for *subjective criteria* involved in the assessment of her behaviour as well?

The answer to the above question is negative: by itself, the fact that an action is compatible with the agent's intentions is not sufficient to conclude that the agent really intends to do that action, in case she performs it. This depends on the fact that, using a famous expression introduced in Anscombe (1957), an individual action can be intentional under one description but not under others. Thus, for example, an instance of intentionally walking to a certain place might be an instance of hitting a bystander; yet, hitting a bystander while walking is something we typically do inadvertently, even when we do it in a way that is compatible with our intentions. In the language of our logic, this means that we cannot use the formula $\langle \alpha \rangle \top$ to single out actions that would be performed intentionally by the agent, if performed by her. For similar reasons, the formula $[1]done(\alpha)$ cannot be used to this end either. Intuitively, this formula says that, if the agent acted in a way compatible with her intentions, then she would necessarily end up doing α. But this can be true in a situation in which the agent does not even know that α is a necessary outcome of acting in accordance with her intentions. For instance, suppose that a doctor wants to give a usually effective cure to a patient, without knowing that this will act as a deadly poison on her. In this case, the doctor cannot act according to her intentions without poisoning the patient. Still, if she does so, we would hardly say that she poisoned the patient intentionally.

So, **AL**, by itself, is still too weak to be a suitable tool to study deontic notions involving the responsibility of the agent and, in particular, the notion of culpable not doing. But let us consider once again scenario 1 in the case study we discussed in section 2. Recall that, in this scenario, John has an excuse for not calling Mr Brown (since his phone is broken), but he does not have any excuse for not going to the law firm. In addition, going to the law firm would be compatible with his intention not to miss the video conference. Now, suppose that John, taking it for granted that his wife will bring the documents to Mr Brown, concludes that going to the law firm is unnecessary. When he decides not to go there, going there becomes incompatible with his intentions. The fact that John *sets himself to omit* to bring the documents to his lawyer thus induces a change in scenario 1, which makes it similar to scenario 2. Crucially, this change is such that, after it occurs, John can be said to *intend* to omit to go to the law firm, so that, if he acts in accordance with his intentions, he omits this action intentionally.

This reasoning suggests that we could account for an agent's intention to omit an action if we could represent the fact that the agent, in any situation, can change her setting or intentions. In fact, after the agent sets herself to omit an action, she can safely be said to intend to omits that action. But how could we improve **AL** so as to model the fact that an agent sets herself to omit an action? One promising and natural way to go is to make the system dynamic in the sense of Public Announcement and Dynamic Epistemic Logics (see Baltag & Moss, 2004; Baltag, Moss, & Solecki, 1998; van Ditmarsch, van der Hoek, & Kooi, 2008; Gerbrandy & Groeneveld, 1997; Plaza, 2007). The rough idea is that a change in the setting of the agent corresponds to an update in the set of individual actions compatible with her setting and, hence, in her overall acting situation. Changes in the agent's setting can thus be modelled as transformations leading from the model representing the agent's original acting situation to the model representing her updated acting situation. The definition of specific update procedures corresponding to the event that an agent sets herself to omit an action requires both a clarification of the concept of omission and a precise characterization of the event that an agent sets herself to omit an action. A full elaboration of this proposal will be presented in currently on-going and future work.

References

Anscombe, G. E. (1957). *Intention*. Oxford: Basil Blackwell.
Baltag, A., & Moss, L. S. (2004). Logics for Epistemic Programs. *Synthese, 139*(2), 165–224.
Baltag, A., Moss, L. S., & Solecki, S. (1998). The Logic of Public Announcements, Common Knowledge, and Private Suspicious. In *Proceedings of the 7th Conference on Theoretical Aspects of Rationality*

and Knowledge (TARK 1998) (pp. 43–56). Morgan Kaufmann Publishers Inc.

Botterell, A. (2009). A Primer on the Distinction between Justification and Excuse. *Philosophy Compass*, *4*(1), 172–196.

Broersen, J. M. (2004). Action Negation and Alternative Reductions for Dynamic Deontic Logics. *Journal of Applied Logic*, *2*(1), 153–168.

Canavotto, I., & Giordani, A. (n.d.). Enriching Deontic Logic. To appear in *Journal of Logic and Computation*.

Castro, P. F., & Maibaum, T. S. (2009). Deontic Action Logic, Atomic Boolean Algebras and Fault-tolerance. *Journal of Applied Logic*, *7*(4), 441–466.

van Ditmarsch, H. P., van der Hoek, W., & Kooi, B. (2008). *Dynamic Epistemic Logic*. Berlin: Springer.

Fletcher, G. P. (1998). *Basic Concepts of Criminal Law*. Oxford University Press.

Gerbrandy, J., & Groeneveld, W. (1997). Reasoning about Information Change. *Journal of Logic, Language and Information*, *6*, 147–169.

Giordani, A., & Canavotto, I. (2016). Basic Action Deontic Logic. In O. Roy, A. Tamminga, & W. Malte (Eds.), *Deontic Logic and Normative Systems* (pp. 80–92). College Publications.

Meyer, J.-J. C. (1988). A Different Approach to Deontic Logic: Deontic Logic Viewed as a Variant of Dynamic Logic. *Notre Dame Journal of Formal Logic*, *29*(1), 109–136.

Plaza, J. (2007). Logics of Public Communications. *Synthese*, *158*(2), 165–179.

Segerberg, K. (1982). A Deontic Logic of Action. *Studia Logica*, *41*(2), 269–282.

Trypuz, R., & Kulicki, P. (2015). On Deontic Action Logics Based on Boolean Algebra. *Journal of Logic and Computation*, *25*(5), 1241–1260.

Vendler, Z. (1957). Verbs and Times. *The Philosophical Review*, *66*(2), 143–160.

von Wright, G. H. (1963). *Norm and Action: A Logical Enquiry*. London: Routledge 7& Kegan Paul.

von Wright, G. H. (1981). On the Logic of Norms and Actions. In R. Hilpinen (Ed.), *New Studies in Deontic Logic: Norms, Actions, and the Foundation of Ethics* (pp. 3–36). Dordrecht: Reidel.

Bases for an Action Logic

Ilaria Canavotto
Universiteit van Amsterdam
Institute for Logic, Language and Computation
The Netherlands
E-mail: i.canavotto@uva.nl

Normality Operators and Classical Collapse

ROBERTO CIUNI AND MASSIMILIANO CARRARA

Abstract: In this paper, we extend the expressive power of the logics K_3, LP and FDE with a *normality operator*, which is able to express whether a formula is assigned a classical truth value or not. We then establish *classical recapture* theorems for the resulting logics. Finally, we compare the approach via normality operator with the classical collapse approach devised by Jc Beall.

Keywords: Many-valued Logic, Classical Recapture, Normality Operators, Classical Collapse, Logic of Formal Inconsistency, Logic of Formal Undeterminedness

Introduction

Theories of *classical recapture* (Beall, 2011, 2013; Priest, 1979, 1991) specify at which conditions we can safely draw *classically valid* inferences while having a (subclassical) *many-valued logic* as our reasoning tool of choice. For instance, if we use Strong Kleene logic K_3 (Kleene, 1952), a theory of classical recapture will specify at which conditions we can assert an instance $\phi \vee \neg \phi$ of the *Law of Excluded Middle*—a principle that fails in the logic. Similarly, if we use the Logic of Paradox LP by Priest (1979, 2006), the theory will specify at which conditions we can apply Modus Ponens and infer ψ from $\phi, \phi \supset \psi$, or apply Ex Contradictione Quodlibet and infer ψ from $\phi \wedge \neg \phi$ — again, the logic fails the rules in question. Something along these lines would be done also for the four-valued FDE (Belnap, 1977).

This endeavor is motivated by a philosophical background that is shared by a number of many-valued logicians: use of a many-valued reasoning tool is necessary because we may face a number of 'abnormal phenomena' that allegedly cannot be treated classically (logical paradoxes, partial information, vagueness, denotational failure), but as long as these phenomena are not at stake, classical logic is perfectly in order as it is.

In this paper, we generalize the expressive power of the *Logic of Formal Inconsistency* (Carnielli, Coniglio, & Marcos, 2007; Carnielli, Marcos,

& De Amo, 2000; da Costa, 1974; Marcos, 2005) and the *Logic of Formal Undeterminedness* (Corbalan, 2012) in order to specify at which conditions we can reason classically when deploying some given many-valued reasoning tools. In particular, we present the many-valued logics $K_3^\circledast$, $LP^\circledast$, and $FDE^\circledast$, which increase the expressive power of K_3, LP and FDE, respectively, and we establish classical recapture results for these logics. Finally, we compare our approach with the classical recapture strategy provided by the *classical collapse* approach by (Beall, 2011, 2013).

The paper proceeds as follows. In the remainder of this Introduction, we provide some background on the Logic of Formal Inconsistency (LFI), the Logic of Formal Undeterminedness (LFU), and the normality operator that we use in the paper. In Section 1, we introduce the logics K_3, LP and FDE, which provide the basic many-valued reasoning tools of the paper. In Section 2, we augment the three logics with the normality operator $\circledast$, thus obtaining systems in the LFI and LFU tradition, and we establish our main results: Theorem 1 and Theorem 2. Interestingly, $FDE^\circledast$ requires a slightly different recapture strategy than $K_3^\circledast$ and $LP^\circledast$. Section 3 introduces the approach by (Beall, 2011, 2013), and Section 4 compares our 'recapture via normality' and classical collapse. Finally, Section 5 summarizes the content of the paper and presents some conclusions.

Background. LFI is a family of systems originating in da Costa (1974). Systems in this family control the behavior of inconsistency by internalizing the notion in the object language. This is done by a *consistency operator*—see end of Section 2. LFI includes a huge variety of formalisms, which may receive highly diversified semantical treatments. Here, we follow Carnielli et al. (2000) in focusing on a formalism that has a straightforward truth-functional semantics (see Sections 1 and 2). LFU dualizes da Costa's project and includes systems controlling the behavior of undeterminedness (failure of Excluded Middle). This is done by an *determinedness* operator that, together with negation, internalizes the notion in the object language.

The *normality operator* from this paper generalizes the operators from LFI and LFU. While the consistency (determinedness) operator expresses that a formula ϕ is consistent (determined), the normality operator expresses the stricter notion that a formula ϕ has a classical truth value (or 'is normal'). While the operators of normality and consistency (determinedness) coincide in a paraconsistent (paracomplete) three-valued logic, they are in principle distinct in a four-valued logic that is both paraconsistent and paracomplete, such as logic $FDE^\circledast$ from Section 2.

Normality Operators and Classical Collapse

1 Preliminaries

Given an infinitely denumerable set $\mathcal{P}$ of propositional variables, standard propositional language $\mathcal{L}_1(\mathcal{P})$ is defined by the following Backus-Naur Form (BNF):

$$\Phi ::= p \mid \neg \phi \mid \phi \vee \psi \mid \phi \wedge \psi$$

where $p \in \mathcal{P}$ and $\neg, \vee, \wedge$ are negation, disjunction, conjunction, respectively. As usual, we define $\phi \supset \psi = \neg \phi \vee \psi$. We denote sets of arbitrary formulas by $\Sigma, \Gamma, \Delta, \ldots$, and we omit reference to $\mathcal{P}$ when possible. We interpret the formulas in $\mathcal{L}_1$ via valuation functions:

Definition 1 (Valuations) *We let $\mathcal{V}$ be the class of all functions $\nu : \Phi \longmapsto \{0, \frac{1}{2}, 1\}$ that satisfy the following clauses:*

- $\nu(\neg \phi)$ $=$ $1 - \nu(\phi)$
- $\nu(\phi \vee \psi)$ $=$ $max(\nu(\phi), \nu(\psi))$
- $\nu(\phi \wedge \psi)$ $=$ $min(\nu(\phi), \nu(\psi))$

We denote by $\mathcal{V}_{CL}$ the set of valuations $\nu \in \mathcal{V}$ such that $\nu(p) \in \{0, 1\}$ for every $p \in \mathcal{P}$.[1] We define a *logic* S semantically as a pair $\langle \mathcal{L}, \vDash_S \rangle$, where $\mathcal{L}$ is a language and $\vDash_S$ is a relation of *logical consequence*—from now on, we will often talk about S-consequence, depending on the system we are focusing on. For every logic S, we define a set $\mathcal{D}_S \subseteq \mathcal{T}$ of *designated values* of S. We define S-consequence as preservation of designated values in S:

Definition 2 (S-consequence) *For every logic S, S-consequence is a relation $\vDash_S \subseteq 2^{\Phi} \times \Phi$ such that:*

$$\Sigma \vDash_S \psi \Leftrightarrow \nu(\psi) \in \mathcal{D}_S \text{ if } \nu(\phi) \in \mathcal{D}_S \text{ for every } \phi \in \Sigma$$

The following is a useful notation: $var(\Sigma)$ is the set of variables p that occur in some $\phi \in \Sigma$. We write $var(\psi)$ instead of $var(\{\psi\})$. We call a *tautology* any formula that follows from the empty set of premises.

1.1 Strong Kleene Logic and the Logic of Paradox

Strong Kleene Logic K_3 (Kleene, 1952) and the Logic of Paradox LP (Priest, 1979, 2006) have found prominent applications in philosophical logic, especially with respect to *logical paradoxes* and *truth theory* (Field, 2008;

---
[1] We believe the reason for the label is clear: the valid rules and principles of Classical Logic CL are determined by these valuations.

Kripke, 1975; Priest, 1979, 2006).[2] Their interpretation is based on the valuation functions from Definition 1. The difference between the two logics is in their designated values: $\mathcal{D}_{K_3} = \{1\}$, while $\mathcal{D}_{LP} = \{\frac{1}{2}, 1\}$.

A straightforward consequence of Definition 1 and $\mathcal{D}_{K_3} = \{1\}$ is that no formula $\phi \in \Phi_{\mathcal{L}_1}$ is a *tautology* in K_3: we have $\nu(\phi) = \frac{1}{2}$ if $\nu(p) = \frac{1}{2}$ for every $p \in var(\phi)$. *A fortiori*, the *Law of Excluded Middle* fails:

$$\emptyset \nvDash_{K_3} \phi \vee \neg\phi \qquad \text{(Failure of LEM)}$$

According to standard terminology, this makes K_3 a *paracomplete* logic. Another consequence of K_3 having no tautology is failure of the *Law of Identity* (LI) $\phi \supset \phi$. Definition 1 and $\mathcal{D}_{LP} = \{\frac{1}{2}, 1\}$ imply that *Ex Contradictione Quodlibet* fails:

$$\phi \wedge \neg\phi \nvDash_{LP} \psi \qquad \text{(Failure of ECQ)}$$

Any $\nu \in \mathcal{V}$ such that $\nu(p) = \frac{1}{2}$ and $\nu(q) = \mathbf{f}$ provides a countermodel. Also, notice that every formula $\phi \in \Phi_{\mathcal{L}_1}$ is *satisfiable* in LP: $\nu(\phi) = \frac{1}{2}$ whenever $\nu(p) = \frac{1}{2}$ for every $p \in var(\phi)$. Finally, all tautologies from Classical Logic CL are LP-tautologies, and vice versa. We refer the reader to Priest (1979) for this.

More in general, presence of the third value implies departure from *classical* consequence $\vDash_{CL}$, to the effect that some classically valid inferences fail in K_3 and LP. The following observation details some validities and the most notable failures of K_3 and LP:[3]

Observation 1 K_3-*consequence and* LP-*consequence satisfy*:

1a	$\psi \nvDash_{K_3} \phi$		1b	$\psi \vDash_{LP} \phi$
	for ϕ a classical tautology			for ϕ a classical tautology
2a	$\phi, \neg\phi \vDash_{K_3} \psi$		2b	$\phi, \neg\phi \nvDash_{LP} \psi$
3a	$\phi \supset (\psi \wedge \neg\psi) \vDash_{K_3} \neg\phi$		3b	$\phi \supset (\psi \wedge \neg\psi) \nvDash_{LP} \neg\phi$
4a	$\phi, \phi \supset \psi \vDash_{K_3} \psi$		4b	$\phi, \phi \supset \psi \nvDash_{LP} \psi$
5a	$\neg\psi, \phi \supset \psi \vDash_{K_3} \neg\phi$		5b	$\neg\psi, \phi \supset \psi \nvDash_{LP} \neg\phi$
6a	$\phi \supset \psi, \psi \supset \zeta \vDash_{K_3} \phi \supset \zeta$		6b	$\phi \supset \psi, \psi \supset \zeta \nvDash_{LP} \phi \supset \zeta$

---

[2] Other applications include *partial functions* (Kleene, 1938, 1952), *partial information* (Abdallah, 1995), *logic programs* (Fitting, 1985) (K_3), and *vagueness* (Priest, 2013; Ripley, 2013; Shapiro, 2006) (K_3 and LP).

[3] We refer the reader to (Beall 2011, 2013; Priest 1979, 2006) for these failures and validities.

Normality Operators and Classical Collapse

As is well known, departure from classical reasoning is the key of K$_3$ and LP's success in approaching the logical paradoxes. However, this success comes at a cost: LP fails Modus Ponens (MP)—failure 4b above—and K$_3$ fails LI, which are crucial in our understanding of a conditional. This suffices to explain why we may want to reason classically when we are sure that no abnormal phenomenon is around. In Sections 2 and Sections 3, we approach two different ways to recapture *classical reasoning* in K$_3$ and LP.

Remark 1 (Reading of the third value) Third value $\frac{1}{2}$ has two natural informal readings in K$_3$ and LP, respectively. Failure of LEM in K$_3$ suggests that $\frac{1}{2}$ is read as 'neither true nor false', or 'undetermined', or 'undefined'. Failure of ECQ in LP suggests that $\frac{1}{2}$ is read as 'both true and false', or 'overdetermined', or 'inconsistent'. These readings will help in what follows.

1.2 First-degree entailment

The logic FDE has been first introduced by Anderson and Belnap (1962), and it has been later generalized to a 'useful four-valued logic' by Belnap (1977). In FDE, formulas from $\mathcal{L}_1$ are interpreted by adjusting Definition 1 to a poset $\{0, \mathbf{n}, \mathbf{b}, 1\}$ of truth values,[4] whose weak partial order $\leq$ is defined as follows:[5]

- $0 < \mathbf{n} < 1$
- $0 < \mathbf{b} < 1$
- $\mathbf{n} \not\leq \mathbf{b}$ and $\mathbf{b} \not\leq \mathbf{n}$

Definition 3 (Valuations, 2) *We let $\mathcal{U}$ be the class of all functions $u : \Phi \longmapsto \{0, \mathbf{n}, \mathbf{b}, 1\}$ that satisfy the following clauses:*

- $u(\neg \phi) = \begin{cases} 1 - u(\phi) & \text{if } u(\phi) \in \{0,1\} \\ x & \text{if } u(\phi) = x \quad \text{for } x \notin \{0,1\} \end{cases}$
- $u(\phi \vee \psi) = glb(u(\phi), u(\psi))$
- $u(\phi \wedge \psi) = lub(u(\phi), u(\psi))$

---

[4]The usual notation for the truth values of FDE is $\mathbf{f}, \mathbf{n}, \mathbf{b}, \mathbf{t}$. However, $\mathbf{f}$ and $\mathbf{t}$ behave exactly as 0 and 1 in K$_3$ and LP, and we keep the numerical notation here, for the sake of uniformity.

[5]A weak partial order is any *reflexive* and *transitive relation* $\mathcal{R}$ on a domain D that obeys $\forall x, y \in D : \mathcal{R}(x,y)$ and $\mathcal{R}(y,x) \Rightarrow x = y$.

Besides, we have $\mathcal{D}_{\mathsf{FDE}} = \{\mathbf{b}, 1\}$. From the specification of the order and the definition of $\mathcal{D}_{\mathsf{FDE}}$, it is easy to see that, if we restrict valuations in $\mathcal{U}$ to $\{0, \mathbf{b}, 1\}$, we obtain LP. Dually, if we restrict valuations in $\mathcal{U}$ to $\{0, \mathbf{n}, 1\}$, we obtain K_3. This suffices to understand that FDE is a sublogic of both K_3 and LP. The following failure guarantees that FDE is a *proper* sublogic of the two formalisms:

$$\phi \wedge \neg\phi \nvDash_{\mathsf{FDE}} \psi \vee \neg\psi \qquad \text{(Failure of Confusion)}$$

2 Recapture via normality

In this section, we propose to recapture classical reasoning by improving the expressive power of K_3 and LP by devices that tell apart the situations where a formula ϕ has a classical truth value ('is normal') from the situations where ϕ has some non-classical value ('is abnormal'). In particular, the logics $\mathsf{K}_3^\circledast$, $\mathsf{LP}^\circledast$ and $\mathsf{FDE}^\circledast$ that we introduce in this section are obtained by extending K_3, LP and FDE, respectively, with a normality operator $\circledast$. The following is a semantic definition of a normality operator:

Definition 4 (Normality operator) *Given a language $\mathcal{L}$, a set $\mathcal{T}$ of truth values including 0 and 1, and valuation functions $v : \Phi_\mathcal{L} \longmapsto \mathcal{T}$, a unary connective k is a normality operator iff, for every $\phi \in \Phi_\mathcal{L}$:*

$$v(k\phi) = 1 \Leftrightarrow v(\phi) \in \{0, 1\} \text{ and } v(k\phi) = 0 \Leftrightarrow v(\phi) \notin \{0, 1\}$$

The logic $\mathsf{LP}^\circledast$ is a LFI along the tradition of da Costa (1974), Carnielli et al. (2007), and Marcos (2005). The logic $\mathsf{K}_3^\circledast$ is a LFU along the lines of Corbalan (2012). We come back to the connections between normality operators and the two families of logics at the end of this section.

2.1 Normality operator

Given an infinitely denumerable set $\mathcal{P}$ of propositional variables, the language $\mathcal{L}_2(\mathcal{P})$ is defined by the following BNF:

$$\Phi ::= p \mid \neg\phi \mid \phi \vee \psi \mid \phi \wedge \psi \mid \circledast\phi$$

where $p \in \mathcal{P}$ and $\circledast$ is a *normality* operator, with $\circledast\phi$ reading 'ϕ has a classical truth value'. We generalize the definition of valuation functions from Section 1:

Normality Operators and Classical Collapse

Definition 5 (Valuations, 3) *We let $\mathcal{V}^+$ be the class of all functions $\nu : \Phi_{\mathcal{L}_2} \longmapsto \{0, \frac{1}{2}, 1\}$ that satisfy the clauses from Definition 1 together with:*

$$\nu(\circledast\phi) = \begin{cases} 1 & \text{if } \nu(\phi) \in \{0,1\} \\ 0 & \text{if } \nu(\phi) \notin \{0,1\} \end{cases}$$

We define $K_3^\circledast = \langle \mathcal{L}_2, \vDash_{K_3^\circledast}\rangle$ as the *extension* of K_3 with $\circledast$. $K_3^\circledast$-consequence $\vDash_{K_3^\circledast}$ is defined according to Definition 2 by assuming $\mathcal{D}_{K_3^\circledast} = \mathcal{D}_{K_3} = \{1\}$. We define $LP^\circledast = \langle \mathcal{L}_2, \vDash_{LP^\circledast}\rangle$ as the *extension* of LP with $\circledast$. $LP^\circledast$-consequence $\vDash_{LP^\circledast}$ is defined according to Definition 2 by assuming $\mathcal{D}_{LP^\circledast} = \mathcal{D}_{LP} = \{\frac{1}{2}, 1\}$.

Given the clause from Definition 5 and $\mathcal{D}_{K_3^\circledast} = \{1\}$, we have $\nu(\circledast\phi) \in \mathcal{D}_{K_3^\circledast}$ iff $\nu(\phi \vee \neg\phi) = 1$: in the logic, $\circledast\phi$ states that ϕ is *determined*—it verifies its corresponding instance of LEM. In turn, this equates with stating that ϕ has a classical value. Dually, given the clause from Definition 5 and $\mathcal{D}_{LP^\circledast} = \{\frac{1}{2}, 1\}$, we have $\nu(\circledast\phi) \in \mathcal{D}_{LP^\circledast}$ iff $\nu(\phi \wedge \neg\phi) = 0$: in the logic, $\circledast\phi$ states that ϕ is *consistent*—it does not satisfy the corresponding contradiction. Again, this means that ϕ has a classical value. The following observation details some validities for the two logics:

Observation 2 $K_3^\circledast$-*consequence and* $LP^\circledast$-*consequence satisfy:*

1a. $\varnothing \vDash_{S^\circledast} \circledast\circledast\phi$ 1b. $\varnothing \vDash_{S^\circledast} \circledast\neg\circledast\phi$
2a. $\circledast\phi \vDash_{K_3^\circledast} \phi \vee \neg\phi$ 2b. $\phi \vee \neg\phi \vDash_{K_3^\circledast} \circledast\phi$
3a. $\neg\circledast\phi \vDash_{LP^\circledast} \phi \wedge \neg\phi$ 3b. $\phi \wedge \neg\phi \vDash_{LP^\circledast} \neg\circledast\phi$
4a. $\neg\circledast\phi \vDash_{S^\circledast} \neg\circledast\neg\phi$ 4b. $\neg\circledast\neg\phi \vDash_{S^\circledast} \neg\circledast\phi$
5a. $\circledast(\phi \wedge \psi) \vDash_{S^\circledast} \circledast\phi \wedge \circledast\psi$ 5b. $\circledast\phi \wedge \circledast\psi \vDash_{S^\circledast} \circledast(\phi \wedge \psi)$
6a. $\circledast(\phi \vee \psi) \vDash_{S^\circledast} \circledast\phi \vee \circledast\psi$ 6b. $\circledast\phi \wedge \circledast\psi \vDash_{S^\circledast} \circledast(\phi \vee \psi)$

where $S^\circledast \in \{K_3^\circledast, LP^\circledast\}$.

In particular, 1a and 1b states that, in $K_3^\circledast$ ($LP^\circledast$), talk about determinedness (consistency) and undeterminedness (inconsistency) are themselves determined (consistent). 2a and 2b state the above equivalence between 'ϕ has a classical truth value' and 'ϕ is determined' in $K_3^\circledast$. Dually, 3a and 3b state the equivalence between 'ϕ does not have a classical truth value' and 'ϕ is inconsistent' in $LP^\circledast$. The remaining items detail how $\circledast$ interacts with the other connectives.

2.2 Normality operators and classical recapture

We need some preliminaries before we establish our classical recapture theorems from $K_3^{\otimes}$, $LP^{\otimes}$ and $FDE^{\otimes}$. First, we need an auxiliary notion:

Definition 6 (Normal counterpart) *Given a set $\Sigma \subseteq \Phi_{\mathcal{L}_1}$, we say that the set $\Sigma^{\otimes} = \{\otimes\phi \in \Phi_{\mathcal{L}_2} \mid \phi \in \Sigma\}$ is the* normal counterpart *of Σ.*

Second, the following observation will prove useful in the next theorem:

Observation 3 *For every $\nu \in \mathcal{V}$, there exists a $\nu' \in \mathcal{V}_{CL}$ such that (1) $\nu'(p) = \nu(p)$ if $\nu(p) \in \{0,1\}$, and (2) $\nu'(p) = 1$ if $\nu(p) = \frac{1}{2}$. By the clauses on ν by Definition 1, it follows that:*

1. $\nu'(\phi) = 0$ if $\nu(\phi) = 0$
2. $\nu'(\phi) = 1$ if $\nu(\phi) = 1$

Now we can establish our *classical recapture theorem*:[6]

Theorem 1 (Recapture via normality) *If $\Sigma, \psi \subseteq \Phi_{\mathcal{L}_1}$, then:*

$$\Sigma \vDash_{CL} \psi \Leftrightarrow \begin{cases} \Sigma, \otimes\psi \vDash_{K_3^{\otimes}} \psi \\ \Sigma, \Sigma^{\otimes} \vDash_{LP^{\otimes}} \psi \end{cases}$$

Proof. We start with $K_3^{\otimes}$: ($\Rightarrow$) Assume $\Sigma \vDash_{CL} \psi$. Now suppose that $\Sigma, \otimes\psi \nvDash_{S^+} \psi$. This implies that there is a $\nu \in \mathcal{V}$ such that $\nu[\Sigma] = \{1\}$ and $\nu(\psi) = 0$, which in turn contradicts the initial assumption.
($\Leftarrow$) Suppose $\Sigma, \otimes\psi \vDash_{K_3^{\otimes}} \psi$. Since $\Sigma, \psi \subseteq \Phi_{\mathcal{L}_1}$, this equates with the fact that, for every $\nu \in \mathcal{V}$, $\nu(\psi) = 1$ if $\nu[\Sigma] = \{1\}$ and $\nu(\psi) \neq \frac{1}{2}$. Thus, the supposition that there is a classical valuation $\nu \in \mathcal{V}_{CL}$ such that $\nu(\psi) = 0$ and $\nu[\Sigma] = \{1\}$ would contradict the initial hypothesis. As as consequence, we have $\nu(\psi) = 1$ for every $\nu \in \mathcal{V}_{CL}$ such that $\nu[\Sigma] = \{1\}$; but this implies $\Sigma \vDash_{CL} \psi$.

As for $LP^{\otimes}$: ($\Rightarrow$) Assume $\Sigma \vDash_{CL} \psi$. If $\mathcal{V}_{CL}(\Sigma) = \emptyset$, then by Observation 3.2, $\{\nu \in \mathcal{V} \mid \nu[\Sigma] = 1\} = \emptyset$. This in turn implies $\Sigma, \Sigma^{\otimes} \vDash_{LP^{\otimes}} \psi$. If $\mathcal{V}_{CL}(\Sigma) \neq \emptyset$, then $\{\nu \in \mathcal{V} \mid \nu[\Sigma] = 1\} \neq \emptyset$. Now suppose that $\nu[\Sigma] = \{1\}$ and $\nu(\psi) = 0$ for some $\nu \in \mathcal{V}$. Take a valuation $\nu' \in \mathcal{V}_{CL}$ such that: (1) $\nu'(p) \in \{0,1\}$ for every $p \in \mathcal{P}$, (2) $\nu'(p) = 1$ if $\nu(p) \in \{\frac{1}{2}, 1\}$, (3) $\nu'[\Sigma] = \{1\}$. By Observation 3.1, $\nu'(\psi) = 0$. Since ν' is classical, this

---
[6] In what follows, we abuse notation a bit and write $\Sigma, \psi \subseteq \Phi$ instead of $\Sigma \cup \{\psi\} \subseteq \Phi$.

contradicts the initial assumption. Thus, $\nu(\psi) = 1$ for every $\nu \in \mathcal{V}_{LP}$ such that $\nu[\Sigma] = \{1\}$. This in turn implies that $\Sigma, \Sigma^{\otimes} \vDash_{LP^{\otimes}} \psi$.

($\Leftarrow$) Suppose $\Sigma, \Sigma^{\otimes} \vDash_{LP^{\otimes}} \psi$. Since $\Sigma, \psi \subseteq \Phi_{\mathcal{L}_1}$, this equates with the fact that $\nu(\psi) \in \{\frac{1}{2}, 1\}$ if $\nu[\Sigma] = \{1\}$. Since $\mathcal{V}_{CL}(\Sigma) \subseteq \{\nu \in \mathcal{V} \mid \nu[\Sigma] = \{1\}\}$, this implies that $\nu(\psi) = 1$ for every $\nu \in \mathcal{V}_{CL}(\Sigma)$. But this implies $\Sigma \vDash_{CL} \psi$. □

Let us see how *recapture via normality* recovers the classical inferences or laws that fail in K_3 and LP—see Section 1:

$$(\phi \wedge \neg\phi), \otimes(\phi \wedge \neg\phi) \vDash_{LP^{\otimes}} \psi \qquad \text{(Recapture of ECQ)}$$

Similarly, LEM and LI can be recaptured in $K_3^{\otimes}$:

$$\otimes\phi \vDash_{K_3^{\otimes}} \phi \vee \neg\phi \qquad \text{(Recapture of LEM)}$$
$$\otimes\phi \vDash_{K_3^{\otimes}} \phi \supset \phi \qquad \text{(Recapture of LI)}$$

More in general, all classical tautologies can be recaptured in $K_3^{\otimes}$ as follows: $\otimes\phi \vDash_{K_3^{\otimes}} \phi$ for ϕ a classical tautology.

Observation 4 $K_3^{\otimes}$-*consequence and* $LP^{\otimes}$-*consequence satisfy:*

$\psi, \otimes\phi \vDash_{K_3^{\otimes}} \phi$ \hfill *for ϕ a classical tautology*

$\phi \supset (\psi \wedge \neg\psi), \otimes(\phi \supset (\psi \wedge \neg\psi)) \vDash_{LP^{\otimes}} \neg\phi$

$\phi, \phi \supset \psi, \{\phi, \phi \supset \psi\}^{\otimes} \vDash_{LP^{\otimes}} \psi$

$\neg\psi, \phi \supset \psi, \{\neg\psi, \phi \supset \psi\}^{\otimes} \vDash_{LP^{\otimes}} \neg\phi$

$\phi \supset \psi, \psi \supset \zeta, \{\phi \supset \psi, \psi \supset \zeta\}^{\otimes} \vDash_{LP^{\otimes}} \phi \supset \zeta$

2.3 Recapture via normality in FDE$^{\otimes}$

In order to develop a 'recapture via normality' strategy for FDE$^{\otimes}$, we consider once again language $\mathcal{L}_2$, which we interpret via the following valuation functions:

Definition 7 (Valuations, 4) *We let $\mathcal{U}^+$ be the class of all functions $u : \Phi_{\mathcal{L}_2} \longmapsto \{0, \mathbf{n}, \mathbf{b}, 1\}$ that satisfy the clauses from Definition 3 together with:*

$$\nu(\otimes\phi) = \begin{cases} 1 & \text{if } \nu(\phi) \in \{0, 1\} \\ 0 & \text{if } \nu(\phi) \notin \{0, 1\} \end{cases}$$

The above definition qualifies FDE$^\otimes$ as the extension of FDE with normality operator $\otimes$. Notice that FDE does not obey Observation 3. In order to see this, take a valuation $u \in \mathcal{U}$ such that $u(p) = \mathbf{n}$, and $u(q) = \mathbf{b}$. Since $glb(\mathbf{n}, \mathbf{b}) = 1$, we have $u((p \wedge \neg p) \vee (q \wedge \neg q)) = 1$. This in turn implies that the recapture recipe from Thoerem 1 does not apply to FDE. Take again the example above, and suppose that, additionally, $u(r) = 0$. Since $u((p \wedge \neg p) \vee (q \wedge \neg q)) = 1$, this suffices to falsify
$(p \wedge \neg p) \vee (q \wedge \neg q), \otimes((p \wedge \neg p) \vee (q \wedge \neg q)), \otimes r \not\models_{\text{FDE}^\otimes} r$. However,
$(p \wedge \neg p) \vee (q \wedge \neg q) \models_{\text{CL}} r$. We deploy a *stronger* recapture strategy, that requires *all* variables from premises and conclusion to have a classical value:

Theorem 2 (Recapture via normality in FDE$^\otimes$) *If* $\Sigma, \psi \subseteq \Phi_{\mathcal{L}_1}$, *then:*

$$\Sigma \models_{\text{CL}} \psi \Leftrightarrow \Sigma, (var(\Sigma))^\otimes, (var(\psi))^\otimes \models_{\text{FDE}^\otimes} \psi$$

Proof. ($\Rightarrow$) Assume $\Sigma \models_{\text{CL}} \psi$. For every valuation $u \in \mathcal{U}$ such that (1) $u[\Sigma] = \{1\}$ and (2) $u(p) \in \{0, 1\}$ for every $p \in var(\Sigma) \cup var(\psi)$, there is a corresponding classical valuation $u' \in \mathcal{U}$. This implies that $u(\psi) = 1$. Otherwise, we would have $u'(\psi) = 0$, which just contradicts the initial hypothesis. Since $\Sigma, \psi \subseteq \Phi_{\mathcal{L}_1}$, this implies $\Sigma, (var(\Sigma))^\otimes, (var(\psi))^\otimes \models_{S^\otimes} \psi$.

($\Leftarrow$) Assume $\Sigma, (var(\Sigma))^\otimes, (var(\psi))^\otimes \models_{\text{FDE}^\otimes} \psi$. Since $\Sigma, \psi \subseteq \Phi_{\mathcal{L}_1}$, this implies that $u(\psi) = \{1\}$ for every $u \in \mathcal{U}$ such that (1) $u[\Sigma] = \{1\}$, and (2) $u(p) \in \{0, 1\}$ for every $p \in var(\Sigma) \cup var(\psi)$. Every such valuation u can be turned into a corresponding classical valuation $u' \in \mathcal{U}$ where, by construction, $u'[\Sigma] = \{1\}$ and $u'(\psi) = 1$. Since these exhaust the classical models of Σ, we can conclude that $\Sigma \models_{\text{CL}} \psi$. □

Notice, however, that the rule of Confusion $\phi \wedge \neg\phi \models \psi \vee \neg\psi$ can be recaptured by just imposing that the formulas in the premises and conclusion have a classical truth value. Indeed, it is easy to check that:

$$\phi \wedge \neg\phi, \{\phi \wedge \neg\phi\}^\otimes, \otimes\psi \models_{\text{FDE}^\otimes} \psi \vee \neg\psi \qquad \text{(Recapture of Confusion)}$$

2.4 Normality operators and logics of formal inconsistency and undeterminedness

The normality operator is a generalization of the consistency and determinedness operators from LFI and LFU, respectively. These families of systems are defined as follows:

Normality Operators and Classical Collapse

Definition 8 (Logic of Formal Inconsistency, Carnielli et al., 2007) *A logic S is a Logic of Formal Inconsistency if and only if it satisfies the following, for some connective k:*

- $\phi, \neg\phi \nvDash_S \psi$ *(Failure of ECQ)*
- $\phi, \neg\phi, k\phi \vDash_S \psi$ *(Principle of Gentle Explosion)*

Definition 9 (Logic of Formal Underminedness, Corbalan, 2012) *A logic S is a Logic of Formal Underminedness if and only if it satisfies the following, for some connective k:*

- $\varnothing \nvDash_S \phi \vee \neg\phi$ *(Failure of LEM)*
- $\varnothing \vDash_S \phi \vee \neg\phi \vee k\phi$ *(Principle of Gentle Implosion)*

Any operator k obeying the criteria from Definition 8 works as a *consistency operator*: the Principle of Gentle Explosion (PGE) implies that $k\phi$ will be designated if and only if ϕ does not verify a contradiction.[7] By contrast, if k obeys the criteria from Definition 9, it will work as a *determinedness operator*: the Principle of Gentle Implosion (PGI) implies that $k\phi$ will be designated if and only if ϕ verifies the corresponding instance of LEM.[8]

This suffices to understand that the normality operator $\circledast$ collapses on a consistency operator in $\mathsf{LP}^{\circledast}$ and any logic that is paraconsistent but not paracomplete. Dually, $\circledast$ collapses on a determinedness operator in $\mathsf{K}_3{}^{\circledast}$ and any logic that is paracomplete but not paraconsistent.

However, $\mathsf{FDE}^{\circledast}$ makes it clear that the normality operator is more general than its kins from LFI and LFU. Definition 4 implies that the normality operator satisfies the criteria from *both* Definition 8 *and* Definition 9 in a logic that is both paraconsistent and paracomplete, like $\mathsf{FDE}^{\circledast}$. By contrast, if we extend FDE with an operator $\circ$ such that $u(\circ\phi) = 1 \Leftrightarrow u(\phi) \neq \mathbf{b}$ and $u(\circ\phi) = 0 \Leftrightarrow u(\phi) = \mathbf{b}$, then we will have a *consistency* operator—that is, an operator satisfying the conditions from Definition 8. Dually, if we extend FDE with an operator $\star$ such that $u(\star\phi) = 1 \Leftrightarrow u(\phi) \neq \mathbf{n}$ and $u(\star\phi) = 0 \Leftrightarrow u(\phi) = \mathbf{n}$, then we will have a *determinedness* operator—that is, an operator satisfying the conditions from Definition 8. Neither of them, however, satisfy Definition 4.

Remark 2 $\mathsf{K}_3{}^{\circledast}$ and $\mathsf{LP}^{\circledast}$ have already appeared in the literature under different names. In particular, $\mathsf{K}_3{}^{\circledast}$ has been first discussed by Gupta and Belnap (1993). As for $\mathsf{LP}^{\circledast}$, this is equivalent with the logic LFI_1 by Carnielli et

---
[7] Notice that what we call recapture of ECQ is equivalent to PGE.
[8] What we call recapture of LEM is equivalent to PGI.

al. (2000). In particular, the latter extends $\mathcal{L}_2$ with a strong negation connective $\sim$ and a detachable conditional $\to$,[9] and it can be obtained from $\mathsf{LP}^\circledast$ by defining $\sim \phi = \circledast \neg \phi$ and $\phi \to \psi = \sim \phi \vee \psi$. On the other hand, $\mathsf{LP}^\circledast$ obtains from LFI_1 since, in that very logic, the consistency operator is definable: $\circledast \phi = \sim \phi \vee \sim\sim \phi$.

3 Classical collapse

Classical collapse is an approach to classical recapture that has been developed by Jc Beall (2011, 2013). This approach is cast against a twofold background: (1) a philosophical view that sees classical logic as our default reasoning tool—a view known as *default classicality*—and (2) a distinction between *logical principles*, which codify a more or less fine-grained reasoning tool, and *extra-logical principles*. The technical results in classical collapse give formal expression to the first background, while the philosophical interpretation of Beall's approach rely on the second background. We start with the formal results.

Default classicality and classical collapse. *Default classicality* is the view that '*classical logic is the default logic, and the weaker logic kicks into when necessary*' (Beall, 2011, p. 326). This view prompts the familiar question— '*How can we* recapture *classical reasoning in our* weaker *logic?*' Beall's reply to it is *classical collapse*.

In order to deploy classical collapse, we need to upgrade standard single-conclusion consequence to multiple-conclusion consequence:

Definition 10 *For every logic* S, S^+-*consequence is a relation* $\vDash_{\mathsf{S}^+} \subseteq 2^\Phi \times 2^\Phi$ *such that:*

$$\Sigma \vDash_{\mathsf{S}^+} \Delta \iff \mathcal{V}_\mathsf{S}(\Sigma) \subseteq \bigcup_{\psi \in \Delta} \mathcal{V}_\mathsf{S}(\psi)$$

with the proviso that, if $\Sigma \vDash_{\mathsf{S}^+} \Delta$, *then* Σ *and* Δ *have finitely many elements.* As usual, if $\Delta = \{\psi\}$, we will write $\Sigma \vDash_{\mathsf{S}^+} \psi$ instead of $\Sigma \vDash_{\mathsf{S}^+} \{\psi\}$.[10] Beside, we need the following:

---

[9]Remember that, in a paraconsistent and not paracomplete logic, $\circledast$ turns to be a *consistency* operator. We follow standard terminology and say that a conditional is detachable if it obeys MP.

[10]Beall (2011, 2013) does not impose the finiteness requirement in its definition of LP^+ and similar multiple-conclusion reasoning tools. However, the restriction is standard, and so we will follow it here.

Normality Operators and Classical Collapse

Definition 11 (Auxiliary definitions)

$$\iota(\Sigma) = \{p \wedge \neg p \in \Phi_{\mathcal{L}} \mid p \in var(\Sigma)\}$$
$$\kappa(\Sigma) = \{p \vee \neg p \in \Phi_{\mathcal{L}} \mid p \in var(\Sigma)\}$$

$\iota(\Sigma)$ is the set of the *contradictions* that can be formed out of the variables from Σ. Thus, if $var(\Sigma) = \{p, q, r\}$, then $\iota(\Sigma) = \{p \wedge \neg p, q \wedge \neg q, r \wedge \neg r\}$. $\kappa(\Sigma)$ is the set of the *instances of LEM* that can be formed from variables from Σ. Again, if $var(\Sigma) = \{p, q, r\}$, then $\kappa(\Sigma) = \{p \vee \neg p, q \vee \neg q, r \vee \neg r\}$. We write $\iota(\psi)$ and $\kappa(\psi)$ instead of $\iota(\{\psi\})$ and $\kappa(\{\psi\})$. Beall (2011,2013) proves the following recapture result for K_3^+ and LP^+:

Proposition 1 (Beall, 2013, Theorem 4.2 and Beall, 2011, Theorem 3.7)

$$\Sigma \vDash_{CL} \Delta \Leftrightarrow \begin{cases} \Sigma, \kappa(\Delta) \vDash_{K_3^+} \Delta \\ \Sigma \vDash_{LP^+} \Delta, \iota(\Sigma) \\ \Sigma, \kappa(\Delta) \vDash_{FDE^+} \Delta, \iota(\Sigma) \end{cases}$$

The proposition states that, when it comes to K_3^+, we can draw the classical conclusions of a given premise-set Σ if *all* variables in the *conclusion* verify *LEM*. Dually, when it comes to LP^+, from a given premise-set we can conclude *either* the classical conclusions of the premise-set, *or* that 'there is something inconsistent in the premise-set.' As for FDE^+, it combines the conditions for the other two logics.

Default classicality as an extra-logical principle. So far so good. But of course, when it comes to LP-reasoning (and its multiple-conclusion version), classical collapse cannot tell us, case by case, *whether we are in an abnormal situation, or in a perfectly classical one*: no formula from $\Phi_{\mathcal{L}}$ can express that 'ϕ has a classical value'. Thus, there is no way for LP (and LP^+) to express *that we are in a normal situation*.

In sum, LP and LP^+ leave us with a *choice* between *classical reasoning* and the *weaker* reasoning tool that is crafted for abnormal situations. In order to make a choice, we need to appeal to *extra-logical* principles (Beall, 2011, p. 331)—principles of rationality, pragmatic principles, epistemic principles, and so on. *Default classicality* can be read as a principle of this sort, stating: *as a first go, reject the inconsistent options* Beall (2011, p. 332)—or, more in general, the *abnormal* options. However, if we face an abnormal case, we switch to the appropriate *weaker* reasoning tool (Beall, 2011, p. 332).[11]

---

[11]This combination of logical and extra-logical principles explains how Beall can insist on

4 Comparing the two approaches

Proposition 1, Theorem 1, and Theorem 2 together imply that *recapture by normality* and *classical collapse* are equivalent:

Corollary 1 *If* $\Sigma, \psi \subseteq \Phi_\mathcal{L}$, *we have:*

$$\Sigma^\oplus, \Sigma \vDash_{\mathsf{LP}^\oplus} \bigvee_{\psi \in \Delta} \psi \quad \Leftrightarrow \quad \Sigma \vDash_{\mathsf{LP}^+} \Delta, \iota(\Sigma)$$
$$\Sigma, \Delta^\oplus \vDash_{\mathsf{K}_3^\oplus} \bigvee_{\psi \in \Delta} \psi \quad \Leftrightarrow \quad \Sigma, \kappa(\Delta) \vDash_{\mathsf{K}_3^+} \Delta$$
$$\Sigma, (var(\Sigma))^\oplus, (var(\Delta))^\oplus \vDash_{\mathsf{FDE}^\oplus} \bigvee_{\psi \in \Delta} \psi \quad \Leftrightarrow \quad \Sigma, \kappa(\Delta) \vDash_{\mathsf{FDE}^+} \Delta, \iota(\Sigma)$$

This is expected: all the methods of classical recapture do exactly the same thing—namely, they establish some kind of *equivalence* with classical reasoning. From a logical point of view, then, the two methods are on a par. However, the two approaches can be compared on a number of extra-logical features.

Semantically closed languages. One virtue of classical collapse over recapture via normality is that Beall's approach can be applied to a semantically closed extension of K_3^+, LP^+ and FDE^+, while recapture via normality cannot apply to semantically closed versions of $\mathsf{K}_3^\oplus$, $\mathsf{LP}^\oplus$, $\mathsf{FDE}^\oplus$. A semantically closed language $\mathcal{L}$ is a language that can express its own concept of truth. This is done by expressions of the form $Tr(\underline{\phi})$, where Tr is a *truth predicate* and $\underline{\phi}$ is the *name* of formula ϕ. Gupta and Belnap (1993) proved that a semantically closed extension of $\mathsf{K}_3^\oplus$ is *trivial*, and Barrio, Pailos, and Szmuc (2016) proved a similar result for $\mathsf{LP}^\oplus$. By contrast, the semantically closed extensions of K_3^+ and LP^+ are not trivial, exactly as those of K_3 and LP.

Informational reading of the many-valued setting. Truth theory and logical paradoxes aren't the only applications of many-valued logic. A consistent track of research applies K_3 and other paracomplete logics to the issue of reasoning with *partial information* (see especially Abdallah, 1995; D'Agostino, 2014; D'Agostino, Finger, & Gabbay, 2013). On the paraconsistent camp, the informational focus is receiving growing attention: Mares (2002); Priest (2001); Restall and Slaney (1995) present theories of *belief revision* that accommodate the presence of inconsistent information.

---

default reasoning without dropping monotonicity: the logic is monotonic, our choices are defeasible (Beall, 2011, p. 332).

Normality Operators and Classical Collapse

If we adopt an informational reading of many-valued reasoning, recapture via normality has some virtues over classical collapse. Indeed, classical collapse for K_3^+ requires the information that *all* variables in the conclusion-set Δ have a classical truth value. From an informational point of view, this is more demanding than the condition provided by Theorem 1 for $K_3^\circledast$, where we just need the conclusion to have a classical value. [12]

As for the paraconsistent case, the classical collapse strategy turns not to be much informative. Indeed, it states what we know from the start: *either we can reason classically, or there is some inconsistency around*. This is not much of a limit *if* classical collapse is complemented by the philosophy of choice in reasoning that is endorsed by Beall (2011, 2013)—indeed, in this case we can opt for the extra-logical principle of *default classicality* and choose one of the two options available. However, the relevance of classical collapse for LP^+ seems to be somewhat smaller out of this philosophical background. By contrast, recapture via normality for $LP^\circledast$ is able to secure that we are in a consistent (classical, normal) situation, thus providing more information than its classical collapse kin. Also, the relevance of this feature does not depend, at least apparently, on a given philosophical background, and it applies to different philosophies of reasoning that one may want to couple with the formal techniques that we have presented in Section 2.

5 Conclusions

In this paper, we have introduced a *normality operator*, which is a linguistic device allowing to distinguish 'normal situations' (where a formula ϕ has a classical value) from 'abnormal situations' (where a formula ϕ has a non-classical value). We have applied the operator in order to build systems of Logic of Formal Inconsistency and Undeterminedness from the many-valued logics K_3, LP and FDE, and we have established classical recapture theorems for the resulting logics $K_3^\circledast, LP^\circledast, FDE^\circledast$ (Theorem 1 and Theorem 2). Finally, we have compared *recapture via normality* with another approach to classical recapture, namely *classical collapse* by Beall (2011, 2013).

---

[12] Notice that, in K_3, some formula can have a classical value even when some of its variables have a non classical one. For instance, $\nu(p \vee q) = 1$ if $\nu(p) = \frac{1}{2}$ and $\nu(q) = 1$.

References

Abdallah, A. (1995). *The Logic of Partial Information*. Berlin: Springer.

Anderson, A., & Belnap, N. (1962). Tautological entailments. *Philosophical Studies*, *13*, 9–24.

Barrio, E., Pailos, F., & Szmuc, D. (2016). A cartography of logics of formal inconsistency and truth. *manuscript*.

Beall, J. (2011). Multiple-conclusion LP and default classicality. *Review of Symbolic Logic*, *4*(2), 326–336.

Beall, J. (2013). LP^+, $K3^+$, FDE^+ and their classical collapse. *Review of Symbolic Logic*, *6*(4), 742–754.

Belnap, N. (1977). A useful four-valued logic. In J. M. Dunn & G. Epstein (Eds.), *Modern uses of multiple-valued logic* (pp. 5–37). Dordrecht: Reidel.

Carnielli, W., Coniglio, M., & Marcos, J. (2007). Logics of Formal Inconsistency. In D. Gabbay & F. Guenthner (Eds.), *Handbook of Philosphical Logic* (Vol. 14, pp. 1–93). Dordrecht: Springer-Verlag.

Carnielli, W., Marcos, J., & De Amo, S. (2000). Formal inconsistency and evolutionary databases. *Logic and Logical Philosophy*, *8*, 115–152.

Corbalan, M. (2012). *Conectivos de Restauracao Local* (Unpublished master's thesis). Universidade Estadual de Campinas.

da Costa, N. C. A. (1974). On the theory of inconsistent formal systems. *Notre Dame Journal of Formal Logic*, *15*, 497–510.

D'Agostino, M. (2014). Analytic inference and the informational meaning of the logical operators. *Logique et Analyse*, *57*, 407–437.

D'Agostino, M., Finger, M., & Gabbay, D. (2013). Semantics and proof-theory of depth-bounded boolean logics. *Theoretical Computer Science*, *480*, 43–68.

Field, H. (2008). *Saving Truth From Paradox*. Oxford University Press.

Fitting, M. (1985). A Kripke-Kleene semantics for logic programs. *The Journal of Logical Programming*, *2*(4), 295–312.

Gupta, A., & Belnap, N. (1993). *The Revision Theory of Truth*. Cambridge, MA: MIT.

Kleene, S. (1938). On notation for ordinal numbers. *Journal of Symbolic Logic*, *3*, 150–155.

Kleene, S. (1952). *Metamathematics*. Amsterdam: North Holland.

Kripke, S. (1975). Outline of a theory of truth. *The Journal of Philosophy*, *72*(19), 690–716.

Marcos, J. (2005). *Logics of Formal Inconsistency* (Unpublished doctoral

dissertation). Universidade Estadual de Campinas.
Mares, E. (2002). A paraconsistent theory of belief revision. *Erkenntnis*, *56*(2), 229–246.
Priest, G. (1979). The logic of paradox. *Journal of Philosophical Logic, 8*, 219–241.
Priest, G. (1991). Minimally inconsistent LP. *Studia Logica, 50*(2), 321–331.
Priest, G. (2001). Paraconsistent belief revision. *Theoria, 67*(3), 214–228.
Priest, G. (2006). *In Contradiction* (2nd ed.). Oxford University Press.
Priest, G. (2013). Vague inclosures. In K. Tanaka, F. Berto, E. Mares, & F. Paoli (Eds.), *Paraconsistency: Logic and Applications* (Vol. 26, pp. 367–378). Berlin: Springer.
Restall, G., & Slaney, J. (1995). Realistic belief revision. In *WOCFAI* (pp. 367–378).
Ripley, D. (2013). Sorting out the sorites. In K. Tanaka, F. Berto, E. Mares, & F. Paoli (Eds.), *Paraconsistency: Logic and Applications* (Vol. 26, pp. 329–348). Berlin: Springer.
Shapiro, S. (2006). *Vagueness in Context*. Oxford University Press.

Roberto Ciuni
University of Padova
Department FISPPA, Section of Philosophy
Italy
E-mail: `roberto.ciuni@unipd.it`

Massimiliano Carrara
University of Padova
Department FISPPA, Section of Philosophy
Italy
E-mail: `massimiliano.carrara@unipd.it`

Reconsidering the 'Ingredients' of Explicit Knowledge

CLAUDIA FERNÁNDEZ-FERNÁNDEZ[1] AND
FERNANDO R. VELÁZQUEZ-QUESADA

Abstract: This paper reviews alternative formal definitions of the concept of *explicit knowledge*, with the goal of unravelling the central notions that different logical approaches employ. The meanings of these notions are classified, explaining how their differences might change the logical outcome. Then, the paper proposes an abstract framework where the most suitable notions come together, with the aim of shedding some light on their underlying theoretical foundations, and also of clarifying their relationship.

Keywords: Knowledge, Awareness, Justification, Actions.

1 Introduction

It is well-known that, in classic epistemic logic (*EL* from now on) (Hintikka, 1962), the agents' knowledge is closed under logical consequence. This property, useful in some applications, is nevertheless an unrealistic assumption when modelling 'real' agents;[2] after all, the purpose of disciplines as Mathematics and Computer Science is to fill in the logical consequences of what we already know. One of the most prominent ideas for 'solving' this logical omniscience problem has been to acknowledge that there are different notions of knowledge (or, more precisely, there are different logical accounts of the notion of *information*; van Benthem & Martínez, 2008). From this perspective, *"the K operator really just describes implicit semantic information of the agent, which definitely has the preceding closure property. The point is rather that closure need not hold for a related, but different intuitive notion [of] explicit [...] knowledge [...], in some suitable sense to be defined"* (van Benthem & Velázquez-Quesada, 2010, p. 6).

---

[1] Partially supported by the Spanish Ministry of Science Project number TIN15-70266-C2-P-1 and the European Regional Fund Development (ERFD).

[2] Or even computational ones, which might lack the required resources (space, time).

Claudia Fernández-Fernández and Fernando R. Velázquez-Quesada

Although this notion of *explicit knowledge* is intuitively clear, different proposals have provided different definitions. For example, while some authors take it as a primitive notion (e.g., Levesque, 1984), others define it as the implicit knowledge the agent is *aware of* (Fagin & Halpern, 1988), some others define it in terms of a notion of *awareness that* (Velázquez-Quesada, 2014), and further ones require an appropriate *justification* (Artemov & Kuznets, 2009). All in all, in different approaches, the concept of *explicit knowledge* is built-up from different 'ingredients'.

This paper starts by reviewing some of the most prominent proposals discussing explicit knowledge (namely, the ones concerning the concepts of *awareness of* and *awareness that*, with the latter related to that of *justification*), identifying and debating the involved notions (Section 2). Then, it proposes an abstract framework where those concepts considered to be most suitable come together. By creating a common setting for the discussed approaches, the proposal sheds some light on the theoretical foundations that underlie them, and thus clarifies their relationship (Section 3). Moreover, it also brings to light the different *epistemic actions* that are considered crucial in each case (Section 4).[3]

2 Main epistemic concepts: the ingredients

An informal and intuitive approach to the notion of "knowledge" will automatically give rise to what is called "explicit knowledge" in formal *EL* developments. "Explicit" will then always refer to the information the agent actually has and is able to access and perform decisions with. In this sense "explicit" stands here for "real" or "actual".

The background intuition that supports the introduction of this notion is the so called "problem of logical omniscience" which therefore involves the concept of "agents with limited reasoning abilities". Here arises the need for modelling a type of knowledge that is not idealized and is applicable to those real agents (represented by human beings and computing machines).

This concept is different from the idealized knowledge that can be found in standard *EL*; which may be better called "implicit". Hence, "implicit knowledge" will always refer to some idealization of knowledge (and correspond to a logical construct).

---

[3] For space reasons, this proposal only discusses the concepts involved in the definition of explicit knowledge (as well as their relationship). A formal counterpart, proposing a semantic structure and a formal language for representing all of them, is left for future work.

Reconsidering the 'Ingredients' of Explicit Knowledge

The distinction between implicit and explicit information is not new. We find one of the first proposals in Levesque (1984); and another relevant one, in Fagin and Halpern (1988), that includes the agent's awareness in their system. We will now dig a bit deeper into these two proposals concerning explicit knowledge in order to find some of the mentioned 'ingredients'.

Deductive system by Konolige (1986) and Levesque (1984): in this approach (Figure 1), explicit knowledge is defined as the primitive knowledge the agent has. Implicit knowledge is then what follows (deductively) from what is explicitly known. Such interpretation assumes that the only action available to the agent is *deductive inference*. It is precisely the possibility of performing this inference what creates the implicit knowledge set.

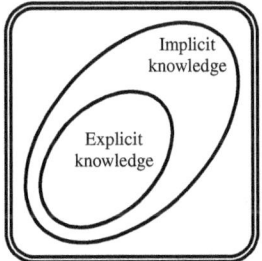

Figure 1: Implicit knowledge and explicit knowledge in Konolige (1984) (diagram from Konolige, 1986).

In the diagram we have a rectangle representing all true propositions and two ellipses representing knowledge. The small one represents those propositions the agent really knows; the large one refers to those propositions she can get to know by deduction.

Awareness Logic by Fagin and Halpern (1988): in their initial proposal of "Awareness Logic" (*AL* from now on), explicit knowledge is defined as the implicit knowledge the agent is aware of. The agent's awareness represents the information she entertains, irrespectively of its condition or the agent's inclination towards it. Note also that what is here called "implicit knowledge" is not what follows from the agent's explicit knowledge, as it was in the previous approach, but rather what the agent would know explicitly if she were aware of every formula that is true in all her epistemic possibilities. The authors take the implicit knowledge from standard *EL* as their starting point. By adding the agent's awareness, which acts as a filter, they obtain explicit knowledge.[4]

---

[4] A similar strategy for defining explicit knowledge is used in epistemic justification logic

In Figure 2 below, we see how implicit knowledge intersects with awareness and gives rise to explicit knowledge.

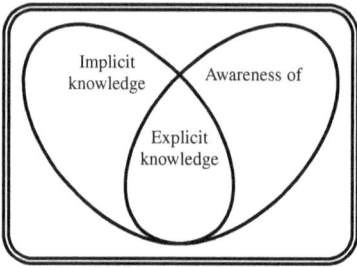

Figure 2: Implicit knowledge, awareness and explicit knowledge in Fagin and Halpern (1988) (diagram from Konolige, 1986).

The concept of "awareness" deserves special mention, since it is a very polysemic term. The authors are (and we will be) always using "awareness" in an epistemic sense, as opposed to a moral sense. But there is still the need for specifying the different uses that the literature on *AL*, and its developments, have made of the term. For doing so, we want to allude to a very useful distinction: *awareness of* vs. *awareness that*.

In Dretske (1993), the author distinguishes "awareness of things" (aware *of* X) from "awareness of facts" (aware *that* X is the case). We could say that, for the author, awareness of facts stands for the awareness of those thoughts (or believed propositions) we can form and ascribe a truth value to; awareness of things would be somehow the very fact of perceiving and forming our mental content, irrespectively of its truth.

Although the difference he proposes stems from a philosophical (and not a logical) point of view, we find it very clarifying. Applying this distinction will result in different interpretations of the awareness operator. In addition, it shows different views of its connection to the implicit knowledge operator when defining explicit knowledge.

Keeping this in mind, we can establish that the awareness Fagin and Halpern are dealing with in their proposal corresponds to Dretske's awareness of things, since they emphasize the fact that awareness can include any type of information. In fact, the two primitive concepts in *AL*, implicit knowledge and awareness, are completely independent of each other.

(Artemov, 2008; Renne, 2012), where explicit beliefs are defined as those implicit beliefs the agent has a justification for.

Reconsidering the 'Ingredients'
of Explicit Knowledge

We could also conclude that in *AL*, explicit knowledge is in some sense analogous to Dretske's awareness of facts, since it corresponds exactly to that part of awareness that intersects with implicit knowledge, and hence is formed only by true propositions the agent is aware of.

Other interpretations of explicit knowledge: different philosophical and logical approaches define what explicit knowledge is and how it can be modelled. But to conclude this part, we want to shed light on one specific sense that can be found in the seminal work of *EL*: Hintikka (1962).

Sillari (2008) calls our attention to an interesting distinction that Hintikka makes in his work: the difference between a weak and a strong sense of "knowing". On the one hand, knowing something *weakly* refers to the fact of entertaining this information and knowing it to be true (analogous to the intuitive sense of explicit knowledge). On the other hand, the *strong* sense of knowing requires not only to be informed about it, but also to have a justification for it.

It is this strong sense of knowing that awakens our interest from both a theoretical and a logical perspective (the weak sense will not be discussed here). We have mentioned that explicit knowledge corresponds to the intuitive sense of "knowing". We could then stretch this informal reasoning and say that for an agent to really explicitly know φ she needs not only to be informed about its truth and be aware of it, but also be able to provide a justification (understanding "justification" as an answer to 'why does she know what she knows?').

This more o less intuitive understanding of real explicit knowledge will be the core of our proposal in the next section. While providing a redefinition of explicit knowledge, we will also classify the other ingredients that have been mentioned above and establish a conceptual framework that allows us to introduce the dynamic actions that transform information.

3 Combining the ingredients

The approaches whose main ideas are depicted in the diagrams of Figures 1 and 2 are not necessarily in conflict with each other. They do look at the logical omniscience problem from different perspectives: one considering agents that might not have 'instantaneous' reasoning abilities (Konolige, 1984), and another considering agents that might not be fully attentive (Fagin & Halpern, 1988). Still, one can combine both ideas in order to create a coherent picture in which there might be different reasons for an agent to not have at her immediate disposal all information she might be able to get.

For this, we take the diagram in Figure 1, which simply distinguishes between the agent's explicit knowledge and its logical consequences (implicit knowledge), as our starting point.[5] Then, we can make a further distinction by bringing awareness into the picture. In doing so, the new ellipse representing awareness will overlap the previous two areas creating two new divisions: the explicit and implicit knowledge the agent is aware of.

In our view, awareness acts as a flashlight that 'illuminates' certain area of the agent's information, making it readily available (reachable) in the sense that the agent can talk about it.[6]

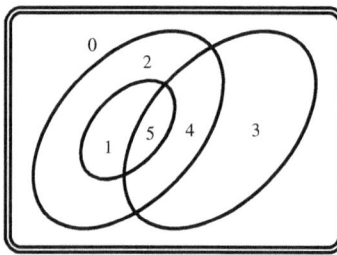

0.- Information

1.- Explicit not in working memory ($\mathbf{K}$)

2.- Implicit not in working memory ($\square$)

3.- Awareness of ($\mathbf{A}$)

4.- Reachable aware knowledge ($\square^{\mathbf{A}}$)

5.- Real explicit knowledge ($\mathbf{K}^{\mathbf{A}}$)

Figure 3: Combined proposal of implicit knowledge, explicit knowledge and awareness.

The diagram depicted above captures the ingredients that have been highlighted in the previous section; it also shows the different types of information a more real agent may have. Let us now devote some words to each one of those zones.

0 - Information: this area corresponds to those propositions outside the agent's current focus. From a theoretical point of view, we could interpret information in a wide sense, standing not only for propositions but also for beliefs, doubts, procedures, etc. But from a formal point of view, we will have to stick only to propositions, as it is done in most logical systems.

---

[5] An analysis similar to the one that follows can also be performed by starting from the diagram in Figure 2 and splitting its *implicit knowledge* area into two, distinguishing in this way what the agent has already derived and what requires further deductive reasoning.

[6] Note: this does not mean that the agent gets to know all the information in the illuminated area; it only makes it part of the current 'topic of conversation', so now the agent can use it (if she happens to know whether it is true or not) or wonder about its truth-value (otherwise).

Reconsidering the 'Ingredients'
of Explicit Knowledge

1 - Explicit not in working memory: in Figure 1, this area (together with 5) corresponds to explicit knowledge. In our diagram, "explicit" stands for the information the agent really knows or has known. Since we distinguish now between what is or not in her working memory (awareness), in this zone 1 we find only those propositions the agent knows for sure, but does not currently have in her working memory. These propositions may be reached by the action of *becoming aware*.

2 - Implicit not in working memory: this zone (together with 4) corresponds to what was called "implicit knowledge" in Figure 1. In this setting, implicit refers to those propositions the agent could reach by deduction (action of *deductive inference*), after becoming aware of some other proposition that permits this inference.

3 - Awareness: the complete ellipse formed by areas 3, 4 and 5 would correspond to the awareness depicted in Figure 2. This part numbered with 3 stands for what we called "awareness of". Here we find everything the agent is currently considering or the questions she is thinking over, but whose truth values cannot be reached by the action of *deductive inference*.

4 - Reachable aware knowledge: this area corresponds to those propositions the agent is already *aware of*, but does not know "really explicitly", since she has not performed the needed inference steps that can provide her with a *justification*. It is a reachable or potential knowledge, in the sense that it is only one deductive step away from being really explicit, that is, from belonging to zone 5.

5 - Real explicit knowledge: lastly, we arrive at the part that has more properties attached to it: the *real explicit knowledge*. As advanced in previous sections, what we refer to here is what Hintikka called "strongly knowing". For the agent to really explicitly know something, she needs to be *aware that* the proposition is true (and therefore aware of) and have a *justification* for it. This justification may come from a deductive inference performed with information she already knew explicitly in this real sense, or by the action of observation, meaning, she was informed about it by an infallible source.

The fact that our real explicit knowledge calls for a justification relies on the intutive claim that 'there is no unjustified knowledge'. But then, this requires a wide-ranged definition of justification that includes not only 'deductive proofs' but also observations. In a more fine-grained approach, we could classify the justifications, and hence the resulting explicit knowledge, depending on the dynamic action(s) that provided the justification.

In the next section, we will discuss some of the dynamic actions that the agent can perform to move information from one zone to another. These actions will highlight not only the dynamics of information, but also the theoretical benefits we obtain when employing this combined diagram.

4 Epistemic actions

The discussion so far has rotated around the 'ingredients' of explicit knowledge: for the agent to really explicitly know a given φ, she must be aware of it and she must have some form of justification for it. This way, we can distinguish between explicit knowledge, what the agent actually has, and different forms of 'obtainable' knowledge, what she might eventually get.

While recognising the different ingredients is useful to create a hierarchy of epistemic notions (some of them without idealised closure properties), doing so also highlights the fact that a given 'piece of information' can be moved across the different hierarchy layers by 'adding' or 'removing' the adequate ingredient. In other words, the agent can move a piece of information across different zones by performing the appropriate *epistemic action*.[7]

Recognising that epistemic actions are involved is crucial. First, providing a list of extra requirements for something to be called explicit knowledge might make the agent non-omniscient, but providing the epistemic actions that allow her to fulfil such requirements guarantees that she will not be defective or ignorant (in other words, she will be *rational*; cf. Duc, 1997). Second, introducing the actions that lead to 'omniscient' states may demystify them. In Conan Doyle's detective stories, the explanation offered at the end turns Holmes' 'magical powers' into a sequence of observations and deductive acts, making the whole procedure "elementary, my dear Watson".

The diagrams in Figures 1 and 2 already hint each at an epistemic action that captures the crucial idea behind the respective proposal (*deductive inference* for the first; *becoming aware* for the second). Both actions still make sense in this proposal's framework, as the following paragraphs describe. Still, its diagram (Figure 3) also allows a systematic analysis that reveals other epistemic actions which are also meaningful in this more general setting.

---

[7] From this perspective, the different forms of 'obtainable knowledge' can be classified in terms of *the actions* the agent needs to perform in order to make the given piece of information really explicit.

Reconsidering the 'Ingredients' of Explicit Knowledge

Deductive inference: this is the crucial action in Konolige (1984) (Figure 1), where implicit knowledge becomes explicit after an act of deductive inference. In our proposal (Figure 3), deductive inference is the action that allows pieces of information to move from region 4 ($\Box^A$: the information the agent is aware of and that follows deductively from what she really explicitly knows) to region 5 (K^A: the information she really explicitly knows).[8]

Becoming aware: this is the crucial action that arises when considering *AL* (Figure 2), where explicit knowledge is the implicit knowledge the agent is aware of. In our proposal (Figure 3), becoming aware stands for different transitions. First, the agent can become *aware of* certain piece of information about which she did not know anything before. This, which can be understood as becoming aware of a brand new possibility, corresponds to a transition from region 0 to region 3 (**A**). But the agent can also become aware of possibilities she does not know anything about, but whose truth can be inferred from her real explicit knowledge; this corresponds to a transition from region 2 ($\Box$) to region 4 ($\Box^A$). Finally, she can also become aware of information she already recognised as true, but did not have 'in her working memory' at the time (e.g., a modal logician who, while watching a football match, is questioned about the finite model property of modal logic); this corresponds to a transition from region 1 (**K**) to region 5 (K^A).[9]

Observation: this action, reflecting intuitively the act of receiving external information, is already meaningful in the individual settings of both discussed papers.[10] It is nowadays called a *public announcement* but, in a single-agent setting, it can be better understood as an act of individual *observation*. By means of it, the agent can turn really explicit (i.e., take to region 5, K^A) not only those pieces of information she entertained but did not know anything about (region 3, **A**), but also those she did not entertain and could not have deduced on her own (region 0), and even those she was unaware of but still could have reached by means of deductive inference (region 2, $\Box$).[11] In fact, an act of observation can move a piece of information from *any* region to region 5 (K^A).[12]

---

[8]Formal epistemic-logic accounts of such action can be found in Ågotnes and Alechina (2007); Duc (1997); Jago (2009); Velázquez-Quesada (2013).

[9]Acts of *becoming aware* have been formally represented and studied in, e.g., van Benthem and Velázquez-Quesada (2010); van Ditmarsch and French (2011); Hill (2010).

[10]It was first formally discussed in Gerbrandy and Groeneveld (1997); Plaza (1989).

[11]Note how the last two possibilities highlight the fact that, by observing a given φ (and thus getting to know that it is the case), the agent also becomes aware of it.

[12]Thus, for example, a 'lazy' agent might know really explicitly both p and $p \to q$, and yet

Further epistemic actions: of course, not all epistemic actions need to lead to a more optimal informational state, and the diagram also makes this clear. Just as one can find actions describing transitions that go 'to the center' of the diagram, it is also possible to find actions that go 'to the corners'. For example, the agent will drop certain piece of information from her 'working memory' if she *becomes unaware* of it.[13] Depending on what further requirements this piece satisfied, this corresponds to either a transition from region 5 ($\mathbf{K^A}$) to region 1 ($\mathbf{K}$), one from 4 ($\square^\mathbf{A}$) to 2 ($\square$), or one from 3 ($\mathbf{A}$) to 0. Another relevant epistemic action is that of *forgetting*[14] which, depending on how the agent acquired the to-be-forgotten information, might take it from region 5 ($\mathbf{K^A}$) to region 4 ($\square^\mathbf{A}$) (she forgets her sister's age, but she can still infer it from her knowledge of her birthday's year), or from region 5 ($\mathbf{K^A}$) to region 3 ($\mathbf{A}$) (she forgets somebody's age, and she does not have enough information to deduce it).

Finally, note how some actions might have 'side effects'. For example, in an appropriate initial state, an observation of $p \wedge q$ can move this formula from region 0 to region 5 ($\mathbf{K^A}$). But this action makes the agent aware of both p and q, and therefore makes any of those formulas 'reachable' via deductive inference (they go from region 0 to region 4, $\square^\mathbf{A}$).

5 Summary and further work

This paper deals with the concept of *explicit knowledge*, commonly used when dealing with non-ideal agents, and corresponding intuitively to what the agent 'currently has'. In its first part it recalls two general strategies for defining this concept, unravelling and exploring the central notions each one employs, and recalling briefly related proposals. In essence, while the first takes explicit knowledge as the *primitive* notion, defining then implicit knowledge as the logical consequences of its explicit counterpart, the second *defines* explicit knowledge as the implicit knowledge the agent is aware of. Therefore, while the first avoids logical omniscience by limiting the agent's (deductive) inferential abilities, the second does it by limiting the concepts she entertains at a given time (i.e., by limiting her current language). This shows how, even though both strategies deal with the same

---

get to know q not by inferring it but rather by observing it.

[13] See, e.g., van Benthem and Velázquez-Quesada (2010); van Ditmarsch, French, and Velázquez-Quesada (2012).

[14] See, e.g., van Ditmarsch, Herzig, Lang, and Marquis (2009); Fernández-Duque, Nepomuceno-Fernández, Sarrión-Morillo, Soler-Toscano, and Velázquez-Quesada (2015).

Reconsidering the 'Ingredients' of Explicit Knowledge

intuitive concept, they follow 'orthogonal' directions, producing then different kinds of agents. Indeed, while agents of the first type have a full language (awareness) but limited reasoning abilities, agents of the second have full inferential abilities within their restricted language (awareness).

In its second part, the paper merges the key ideas behind the discussed frameworks. The resulting setting allows for finer epistemic concepts, as it combines the two 'ways' in which the agent might 'miss' something. Moreover; by doing it, it also highlights the different epistemic *actions* that are involved. Indeed, the combined setting distinguishes not only between the 'truths' the agent has acknowledged and are currently in her 'working memory' (K^A; region 5) and those she has acknowledged but are 'out of the topic of conversation' (K; region 1), but also between those she does not know but might get to know by means of deductive inference ($\square^A$; region 4) and those she does not know but might get to know by means of a raise in awareness *and then* a deductive inference ($\square$; region 2). Finally, it also distinguishes between those truths the agent cannot reach by deductive inference but are still currently entertained (A; region 3) and those that are out of deductive reach and also not being currently 'discussed' (region 0). Thus, on the one hand, the setting has a single notion of 'real' explicit knowledge, corresponding to what the agent has 'in her hand' at the given moment and thus does not require any further epistemic action to be available. On the other hand, it has several notions of obtainable knowledge, each one of them corresponding roughly to the different *sequence of actions* that are needed to turn them explicit.

This paper's goal has been to clarify the concept of explicit knowledge, and the presented discussion of different frameworks dealing with it is a first step. Still, there is further work to be done. One of the most appealing tasks is to propose a formal framework in which all these epistemic notions have a place. An already explored possibility is the use of the awareness logic setting to represent both the agent's language and what she has acknowledged as true (Grossi & Velázquez-Quesada, 2015); this uses implicit knowledge, awareness and 'acknowledgement' as the primitive concepts, defining the rest in terms of them, and follows the *dynamic epistemic logic* framework (van Benthem, 2011; van Ditmarsch, van der Hoek, & Kooi, 2008) for dealing with the involved actions. Still, there are other possibilities. For example, one can use *neighbourhood models* (Montague, 1970; Scott, 1970), under which the primitive concept would be that of explicit knowledge, with its implicit counterpart being definable as the fixed point of a 'deductive closure' operation (Velázquez-Quesada, 2013). One can also follow different directions for modelling awareness: the syntactic approach of Fagin and Halpern (1988) is one possibility, but one can also use semantic tools, as the relational approach used for representing a notion of 'issues being discussed' in logics for questions (i.e., erotetic logics) (Baltag, Boddy, & Smets, 2017; van Benthem & Minică, 2012).

References

Ågotnes, T., & Alechina, N. (2007). The dynamics of syntactic knowledge. *Journal of Logic and Computation*, *17*(1), 83–116.

Artemov, S. N. (2008). The logic of justification. *The Review of Symbolic Logic*, *1*(4), 477–513.

Artemov, S. N., & Kuznets, R. (2009). Logical omniscience as a computational complexity problem. In A. Heifetz (Ed.), *Proceedings of TARK-2009, Stanford, CA, USA, July 6-8, 2009* (pp. 14–23).

Baltag, A., Boddy, R., & Smets, S. (2017). Group knowledge in interrogative epistemology. In H. van Ditmarsch & G. Sandu (Eds.), *Jaakko Hintikka*. Springer.

van Benthem, J. (2011). *Logical Dynamics of Information and Interaction*. Cambridge University Press.

van Benthem, J., & Martínez, M. (2008). The stories of logic and information. In P. Adriaans & J. van Benthem (Eds.), *Philosophy of Information* (Vol. 8, pp. 217–280). Amsterdam: North-Holland.

van Benthem, J., & Minică, S. (2012). Toward a dynamic logic of questions. *Journal of Philosophical Logic*, *41*(4), 633–669.

van Benthem, J., & Velázquez-Quesada, F. R. (2010). The dynamics of awareness. *Synthese*, *177*(Supplement-1), 5–27.

van Ditmarsch, H., & French, T. (2011). Becoming aware of propositional variables. In M. Banerjee & A. Seth (Eds.), *Logic and Its Applications - 4th Indian Conference, ICLA 2011, Delhi, India, January 5-11, 2011. Proceedings* (Vol. 6521, pp. 204–218). Springer.

van Ditmarsch, H., French, T., & Velázquez-Quesada, F. R. (2012). Action models for knowledge and awareness. In W. van der Hoek, L. Padgham, V. Conitzer, & M. Winikoff (Eds.), *International Conference on Autonomous Agents and Multiagent Systems, AAMAS 2012, Valencia, Spain, June 4-8, 2012 (3 Volumes)* (pp. 1091–1098). IFAAMAS.

Reconsidering the 'Ingredients' of Explicit Knowledge

van Ditmarsch, H., Herzig, A., Lang, J., & Marquis, P. (2009). Introspective forgetting. *Synthese, 169*(2), 405–423.

van Ditmarsch, H., van der Hoek, W., & Kooi, B. (2008). *Dynamic Epistemic Logic* (Vol. 337). Dordrecht, The Netherlands: Springer.

Dretske, F. (1993). Conscious experience. *Mind, 102*(406), 263–283.

Duc, H. N. (1997). Reasoning about rational, but not logically omniscient, agents. *Journal of Logic and Computation, 7*(5), 633–648.

Fagin, R., & Halpern, J. Y. (1988). Belief, awareness, and limited reasoning. *Artificial Intelligence, 34*(1), 39–76.

Fernández-Duque, D., Nepomuceno-Fernández, Á., Sarrión-Morillo, E., Soler-Toscano, F., & Velázquez-Quesada, F. R. (2015). Forgetting complex propositions. *Logic Journal of the IGPL, 23*(6), 942–965.

Gerbrandy, J., & Groeneveld, W. (1997). Reasoning about information change. *Journal of Logic, Language and Information, 6*(2), 147–169.

Grossi, D., & Velázquez-Quesada, F. R. (2015). Syntactic awareness in logical dynamics. *Synthese, 192*(12), 4071–4105.

Hill, B. (2010). Awareness dynamics. *Journal of Philosophical Logic, 39*(2), 113–137.

Hintikka, J. (1962). *Knowledge and Belief.* Ithaca, N.Y.: Cornell University Press.

Jago, M. (2009). Epistemic logic for rule-based agents. *Journal of Logic, Language and Information, 18*(1), 131–158.

Konolige, K. (1984). *A Deduction Model of Belief and its Logics* (Unpublished doctoral dissertation). Computer Science Department, Stanford University, Stanford, USA.

Konolige, K. (1986). What awareness isn't: A sentential view of implicit and explicit belief. In J. Y. Halpern (Ed.), *Proceedings of TARK-1986, Monterey, CA, March 1986* (pp. 241–250). Morgan Kaufmann.

Levesque, H. J. (1984). A logic of implicit and explicit belief. In R. J. Brachman (Ed.), *Proceedings of the National Conference on Artificial Intelligence. Austin, TX, August 6-10, 1984.* (pp. 198–202). AAAI Press.

Montague, R. (1970). Universal grammar. *Theoria, 36*(3), 373–398.

Plaza, J. A. (1989). Logics of public communications. In M. L. Emrich, M. S. Pfeifer, M. Hadzikadic, & Z. W. Ras (Eds.), *Proceedings of the 4th International Symposium on Methodologies for Intelligent Systems* (pp. 201–216). Tennessee, USA: Oak Ridge National Laboratory, ORNL/DSRD-24.

Renne, B. (2012). Multi-agent justification logic: Communication and evidence elimination. *Synthese, 185*(Supplement 1), 43–82.

Scott, D. (1970). Advice on modal logic. In K. Lambert (Ed.), *Philosophical Problems in Logic* (pp. 143–173). Dordrecht, The Netherlands: Reidel.

Sillari, G. (2008). Models of awareness. In G. Bonanno, W. van der Hoek, & M. Wooldridge (Eds.), *Logic and the Foundations of Game and Decision Theory (LOFT7)* (pp. 209–240). Amsterdam, The Netherlands: Amsterdam University Press.

Velázquez-Quesada, F. R. (2013). Explicit and implicit knowledge in neighbourhood models. In D. Grossi, O. Roy, & H. Huang (Eds.), *Logic, Rationality, and Interaction - 4th International Workshop, LORI 2013, Hangzhou, China, October 9-12, 2013, Proceedings* (Vol. 8196, pp. 239–252). Springer.

Velázquez-Quesada, F. R. (2014). Dynamic epistemic logic for implicit and explicit beliefs. *Journal of Logic, Language and Information*, *23*(2), 107–140.

Claudia Fernández-Fernández
University of Málaga
Spain
E-mail: `cffernandez@uma.es`

Fernando R. Velázquez-Quesada
Universiteit van Amsterdam
Institute for Logic, Language and Computation
The Netherlands
E-mail: `F.R.VelazquezQuesada@uva.nl`

On What Counts as a Translation

ALFREDO ROQUE FREIRE[1]

Abstract: In this article, instead of taking a particular method as translation, we ask: what does one expect to do with a translation? The answer to this question will reveal, though, that none of the first order methods are capable of fully represent the required transference of ontological commitments. Lastly, we will show that this view on translation enlarge considerably the scope of translatable, and, therefore, ontologically comparable theories.

Keywords: Translation, Ontology, Ontological reduction

1 Relativity of translation

Some fundamental indeterminacies in largely used philosophical concepts were introduced by Quine in his well-known articles *Two Dogmas of Empiricism* (TD) (Quine, 2000) and *Ontological Relativity* (ORel) (Quine, 1968) and in the book *Word and Object* (WObj) (Quine, Churchland, & Føllesdal, 2013). In the first, he makes a resounding critique of the traditional distinction between synthetic and analytic sentences; from this, he concentrates on asserting the indeterminacy of translations between any two languages in the *radical translation* experiment (WObj). In the second, he shows that the absence of distinction, together with the indeterminacy of the models in Löwenheim-Skolem's theorem, implies the inscrutability of the reference relation. Thus, every reference relation is fundamentally linked to the choice of a background theory for which the existential requirements of the theory are interpreted.

If, in TD and WObj, Quine describes the indeterminacy of the translations at the epistemological level, in ORel he relativizes the ontology of a theory to a translation in the background theory: "Specifying the universe of a theory makes sense only relative to some background theory, and only relative to some choice of a manual of translation of the one theory

---

[1] I am grateful to FAPESP for funding this research and Walter Carnielli for the institutional support. Also, I owe much of the discussion in this article to the contributions of Rodrigo Freire, Edgar Almeida, Bruno Ramos, Henrique Antunes and Gilson Olegário.

into the other" (Quine, 1968). Offering an ontology to a theory, therefore, amounts to reducing a theory in a background theory. The translations are, for Quine, underdetermined by any empirical experiment, but not by a kind of relativity.

Quine used the notion of model-theoretic reduction between first-order theories to show the relativity of the reference relation. But the reduction itself would not be subject to the same kind of relativity. If the translation relation has a constitutive defect, then it would be contextual - but not underdetermined. According to Quine, the translation relationship could be reconstructed from the syntactic feature he termed "proxy function". Non-relativity, in this case, is linked to the understanding that translations are not properly objects, that such functions "need not exist as an object in the universe even of a background theory"(Quine, 1968). To take such an exit, however, is as much restricted on the scope of translations as it is invariably interpreted with respect to the mechanisms performed by the background theory. It is restricted because translations not only occur between a background theory and an object theory, but also between two unfamiliar theories to the underlying theory. It is interpreted because a first-order theory is not able to capture the claim that a given set of formulas represent-the-T-theory.

It is only possible to understand what was done by the background theory as a translation if we interpret the result outside the scope of the background theory. If PA is internalized in ZFC, this means that (1) for each formula of PA's language, there exists a set in ZFC that represents the formula of PA, (2) for each sequence of PA's formulas exists a set representing the sequence of PA's formulas and finally (3) that if a sequence represents a PA's proof, then ZFC proves that the set representing the sequence satisfies a definable property in ZFC. All these statements are not statements of ZFC, but of a metatheory that establishes the bond between the two theories. From the ZFC's point of view, one simply proves theorems about certain sets, and the statement that these proofs speak about PA must be interpreted in another metatheory. When Gödel does the internalization of arithmetic in arithmetic itself, he does not assume to be internalizing properly the arithmetic; rather, it is only possible to understand the internalization from the theorem of the representation which is done in primitive recursive Arithmetic (ARP). Eventually, this means that the relationship of internalization only makes sense if we consider a third theory seeing the two arithmetics. Something similar occurs in the case of ZFC interpreting PA: we must take into account at least one ARP that is responsible for establishing that the representation of PA in ZFC is a representation of PA - otherwise, it is not possible to understand in this procedure more than ZFC proving interesting theorems.

On What Counts as a Translation

More generally, we reinforce the idea that any translation relationship between theories can only be understood in a third theory. This means that the case in which we seemingly reduced a theory T in a background theory T_m is, in fact, the evaluation of the relationship between T and T_m in a third theory T_M and the reduction provided by Gödel allows us no more than to say that T_m and T_M can be syntactically the same.

This issue becomes even more evident if we consider the environment of independence proofs in set theory. When, by abbreviation and convenience, we call the Gödel's constructibles L a model for ZFC, we do no more than actually state that there is an *interpretation* (Shoenfield, 1967, p. 57 - 61) between the two theories. In this case, it is not possible to do a complete internalization in the same sense applied in the cases already mentioned. If, in the earlier cases, internalization would provide us with a truth predicate for the internalized theory, the same is not valid for these set theories - if consistent, none of them is capable of internalizing one truth predicate into the other. Notably, the background theory (ARP) for the interpretation is not able to offer an ontology even relative for set theories, and yet it is able to speak about the "translation" between them.

The question still remains open: (QEx) who asserts that the syntactic resources stated in the third theory are a translation relation? Indeed, this seems a new question of representation, in a sense similar to that of the question in the case of reduction, although more serious. As Boghossian (1996) recalls, Quine himself in TD examines two distinct types of analyticity: the first in which the substitution of synonymous terms for forming logical truths is based on the synonymy between a term T and a term T' introduced by definition; and the second in which the basis is a synonymous relation intuited by a competent speaker. In the first case, the relation of synonymy is trivial and of little influence in Quine's critic, whereas the second one is the focus of harsh criticisms. Indeed, QEx is a question whose answer suffers from the same problems as the empiricist dogmas. If we want to analyze translations so that they do not suffer from such criticisms, we must deny them the status of a question capable of expressing meaning and - (1) assume that the two theories analyzed in the translation were defined in the context of the background theory T_B and (2) assume that "being a translation" is a theoretically defined element in T_B. To assume (1) is excessively convenient, though hardly problematic; to assume (2) is artificial. Let's look at these two hypotheses in more detail.

The analysis of the assumption (1) starts with the question of what it means to take a given theory as a metatheory or background theory. That question is not frequently raised because it is obvious or by negligence. When, for example, we study the relationship of satisfaction in ZFC, through model theory, we do much more than using ZFC as the background theory. Before even evaluating the satisfaction of a T theory, one must internalize T's syntax into ZFC; and the fact that this internalization represents T is a statement from another background theory. The notion that an internalized theory "is a theory" must be a ZFC predicate whose (2') representability is guaranteed in a similar fashion to the assumption (2). Even if, again, we take the assumption (2') as unproblematic, we would have to admit that it is not possible to understand a theory that was not internalized in the first place. This means that if an agent A_1 has T_M as background theory and a second A_2 agent presents a theory T_x unknown by A_1, then A_1 cannot offer any understanding about T_x, since he or she could not present a representation predicate "this internalization represents T_x".[2]

Let us assume problem (1) as solved and proceed with the question pointed out in supposition (2). As we have seen, the translation must be defined as a predicate in a background theory that establishes by theorem the link between two theories. This necessity imposes precisely what Quine wanted to avoid concerning translations, i.e. that they were objects. In this case, it is relatively simple to show that what counts as a translation is relative to the choice of a meta-metatheory. Consider, for instance, the case of interpretation between PA and ZF without the axiom of infinity and with the addition of the negation of the axiom of infinity (ZF^{-Inf}). Explicitly, this result can be described as a proof in ARP that if $Th_T(\alpha)$ is the predicate internalized in the ARP that states that α is theorem of theory T and being I the interpretation of PA into ZF^{-Inf}, then

$$ARP \vdash \forall \alpha \in \mathcal{L}_{PA}(Th_{PA}(\alpha) \to Th_{ZF^{-Inf}}(\alpha^I)) \quad (1)$$

We take an undecidable formula δ of PA. If an ARP model (in this case, relativized to the meta-metatheory) satisfies $Th_{PA}(\delta)$, then the number representing the proof must be a non-standard number. Therefore, it is possible to build an ARP model such that $Th_{PA}(\delta)$ is true, while δ is false - say this

---

[2]This problem of representation is not easily solved without extrapolating the first-order environment. I endorse, in this sense, the thesis defended by Freire (2017) that the identity of a mathematical theory is not a formal system, but a normativity instituted by the practice. In the case of a purely formal theory such as the T_x, I add that its identity is instituted in practice with formal systems. This movement toward normativity allows the communication between agents to be the exchange of information at the normative level and the representation of T_x for A_1 is given by the same relation that A_1 already has with the theories of A_1's known scope.

model is $\mathcal{M}$. Notably, for $\mathcal{M}$, the formula α^I is true in ZF^{-Inf} and, at the same time, it is a translation of the formula α of PA. Still, the model sees this same formula as false. Then we come across the strange situation in which a false formula is translated into a true formula. In this case, the interpretation I does not count as a translation relative to the choice of $\mathcal{M}$.

Even so, one could insist on contextualization, stating that the translation of a theory T_1 into another theory T_2 makes sense only when the ontology is fixed for each of the theories. That is, we would think the translation between two theories in the context in which all their sentences were decided by the stipulation of the models $\mathcal{M}_1$ and $\mathcal{M}_2$. Although we take this as an error in establishing priority, we will consider it for the moment. In the article *Satisfaction is not absolute*, Hamkins and Yang (2014) show that two models for ZFC can agree on what the standard model for PA is, and still disagree on which formulas the standard model satisfies. This means that the decision on meta-metatheory can determine the truth value of certain formulas of the model of T_1 or T_2. If, therefore, we translate a formula with this property into another theory, we would not know if we should map it to a true or false sentence in the other theory until we fix the meta-metatheory. Thus, even if the context of the background theory is fixed as a condition for the translation, it is still underdetermined in relation to the meta-metatheory.

But, as stated, the problem is rather one of priority. Offering an ontology to a theory is more complex than offering a translation between two theories - evidence for this is the fact that one can speak much of the translation between two theories without having to touch the subject of an ontology. The ontology for a theory is the answer to the question "what does the theory commit to?", While translation is the answer to the question "how can one theory offer an understanding for the commitment of another theory?". We can have a more understandable and consensual answer to the question "what is the translation of unicorn in Japanese?" before we have an answer to the question "are there unicorns?" - and to say that the answer to the first depends on the response to the second seems unreasonable.

Therefore, translations are relations between two theories established in a third theory. And a theoretical environment must be responsible for asserting that the mapping in question preserves what one wants to preserve. Because of that, as we have seen, the relativity of translation takes effect. Accepting this thesis, however, is not possible in Quine's program, since a significant part of this relies on the concept of ontological reduction.

2 Translation idealized

One way to restore the treatment of the philosophy of mathematics and physics after the attack on meaning is to attribute legitimacy to the ontological reductions between theories. It is for this reason that Quine emphasizes the treatment of ontological reductions. However, we have shown that translations are also subject to the same kind of relativity as the notion of meaning. I understand that Quine himself would impose a limit on his relativism if he came to that conclusion. Despite this, I am still sympathetic to his inquiries - though not to his conclusions. Therefore, we review Quine's implications for meaning, analyticity, and, by extension, for translations.

In *On what there is* (Quine, 1948), Quine establishes the concept of ontological commitment of first-order theories. We would be committed to the existence of entities capable of assuming the role of the bound variables in the axioms, making them true. Further on, in ORel, Quine shows that we are unable to present a determinate ontology to the criterion of ontological commitment in an absolute sense, and so we are forced to speak of ontology in a relative sense. Notably, Quine first asks about what "existence must accomplish" and then shows that we can only relatively accomplish such a requirement. However, for him, the same does not occur with the concept of translation: translations are simply defined as a tool that preserves truth and boolean structure in a very particular way. It is therefore necessary to rekindle the question of what counts as a translation by introducing the question: "What should a translation do?" Or "What does someone who performs a translation expect to do?"

To properly formulate this question, we correlate with the question about existence. If, in the case of existence, we want to know what the statement of the theory requires it to exist, in the case of translation we want to replace the language in such a way that the requirement of existence is preserved. So we start from the definition: With a translation of a theory T_1 into another theory T_2 it is expected that **what T_1 is committed to exist** be transferred to **what T_2 is committed to exist**.

With this definition, we want to emphasize that the relation between the concept of translation and the mappings we carry out in the first order theories is the relation of *formalization*. Much of what is observed in the literature of the subject is the assumption that translations carry out the transport of existential requirements and the assertion of translation between two theories guarantees the reducibility of one ontology into the other. We consider this an error - an exaggerated transparency as to what it may or may not

On What Counts as a Translation

count as a translation. We now turn to a more accurate analysis on what this question can offer us.

We will take a step back, trying to understand what would possibly be an *ideal translation*. We do not want to establish a methodology that obtains ideal relations of translation, but to affirm some necessary (not sufficient) properties so that a method can imply a translation in its maximum sense. Nor do we want to say that this is a translation to be sought, on the contrary, we want to use this abstract experiment to reinforce a subtle problem, namely, that satisfying what is desired with an idealized translation is not at all trivial.

Initially and (in my view) without prejudice, we consider a rather simplified conceptual scheme, in which the defined names and descriptions have only reference relations to objects in the world. We will avoid the problem of the *radical translation* pointed out in TD and WOb, assuming the existence of an ideal mediator (IM). Quine showed that it is necessary for two speakers to have equivalent linguistic/conceptual structures so that they can establish any effective communication about a translation between the theories used by each of them. This IM is able to understand both languages that one wants to translate. In this case, she will be responsible for ensuring that speakers actually refer to the same objects when they, in fact, perform a correct defined description that replaces the description used by another speaker in their own language. Assuming the IM as our mediator, we avoid the epistemological problem and focus on what is ideally a translation.[3]

An important aspect that makes this type of translation possible is: we are assuming as fixed the object of the references. At least in principle, there is a non-linguistic way of accessing object, either by sight, hearing, touch or any indirect way of capturing that same information. If a speaker describes "the stone", we can see, hear or touch that referent. In the case of a description of the type "the mayor of the city", identifying this referent can occur through obtuse and complex ways; if one of the languages does not have a single predicate that is equivalent to "being a mayor", one could still use a predicate like "head of X". The reference of "head of X" still depends on the concept of "citizens of Y" and this concept must be translated in a progressive nesting of attempts and errors until an effective translation can be reached.

---

[3] In view of these considerations, it seems possible to establish a translation between two speakers of two different languages. When one of them, A, refers to an object by a defined description a_1, the other speaker, B, could, by trial and error, finally hit the equivalent defined description b_1 in its own language. Ideally, A and B could begin to seek more complex levels of language until an effective translation is established.

Indeed, we have dealt only with the formulas that have a reference to ensure understanding. However, this does not include all cases of translation, we can, as we well do, translate sentences that have no referents, as the sentence "the king of France is bald". Although we take it simply as false, we cannot say that this sentence has no translation in another language. We can say that the translation preserves the sense of Frege or the positivist's method of verification; independently of this, the sentence, although without reference, has a "potential reference" and this must be preserved. If France is to have a king again and this king is bald, the sentence in both languages must cease to have false truth-value or " be meaningless " and become true. This does not mean a denial of Quine's holism of meaning - in fact, it is perfectly acceptable to admit that, rather than simply accepting that the sentence becomes true, another theoretical assumption is reviewed. It is enough that both speakers are sensitive to the truth-value change of the proposition in question. And if it is the case that this change entails the understanding (through IM) that the speakers are not using the same referent, then this will adjust in the next iterations with IM.

It is not the case that the simple mapping of the formulas with reference must preserve the truth-value, this only guarantees that the translation "works" for the particular experiential universe that the speakers live at the moment. In order for there to be a translation in the strongest sense, it is necessary (and reasonable) that languages preserve translation even if reality changes or if there is any new discovery of the sciences. Therefore, the ideal translation must be able to fix the reference relations, the arrows that link the names to the objects.

We have considered only theories that make direct reference to objects of the world. There is much controversy whether such theories would even be possible. We assume this as possible only as an abstract experiment. This will not influence the discussion, since ultimately we want to talk about theories that do not make direct reference as theories of the first order.

The ideal translation between two theories of **direct reference** T_1 and T_2 should be a mapping between the formulas of T_2 in T_1 such that the existential requirement of a formula of T_2 is the same as the existential requirement of the correspondent in T_1. In the case of theories that do not make direct reference, the picture changes. The terms of the theory do not point directly to the world, but only require that those objects captured by the quantifiers satisfy a certain set of properties. In these cases, we no longer require that a translated sentence point to the same object as the original sentence - it is only necessary that the collection of objects pointed out by both theories be isomorphic.

On What Counts as a Translation

A first order theory does not fix the references, but just how each object captured by the quantifiers relates to each other.[4] That is why, when we speak of first order theories, we affirm reference relations to collections of objects "under isomorphism". In order to have a soundness criterion for the translation, we need to have a criterion to say that the theories refer to the "same" object as we had with the IM. Given a possible reference model $\mathcal{R}_1$ for the theory T_1 and $\mathcal{R}_2$ for the theory T_2, then the reference r_1 of a defined description $d^{trans(T_1)}$ is the same as the reference r_2 of the original description d in T_2 if, and only if, r_1 relates to all other objects translated into $\mathcal{R}_1$ in the same way as r_2 relates to all other objects of $\mathcal{R}_2$.

Yet, why, in the case of theories that do not fix the reference, should we maintain that translation preserves meaning? Given a defined description $d(x)$, we can not know if the description has a reference until the theory proves that $\exists x d(x)$. However, even if we do not know whether this is the case or not, the translation must be able to establish this "transfer of existential requirement". Indeed, some first-order theories are decidable, and for this reason there is a procedure that determines whether or not the predicate refers. For cases of translation between decidable theories, it is arguably valid to assume that the translation need not focus on fixing the reference relations, but only that all descriptions are mapped isomorphically. However, this is not valid in any case, if a theory is incomplete, there are descriptions $d(x)$ we cannot know whether they have reference or not - in this case, we must preserve the existential requirement in the same way that we must preserve the translation for the "king of France" even though he does not exist at the moment. Nevertheless, we now stop talking about "fixing the reference relation" and we proceed to say that the translation must "fix the reference relation in an isomorphic context".

In order to be able to assert an "equality of meaning" criterion independent of the models fixed for the theories, it would be necessary for the descriptions $d^{Trans(T_1)}$ and d to have the same reference for any two models $\mathcal{R}_1$ and $\mathcal{R}_2$ of T_1 and T_2. This criterion, however, cannot be applyed unrestricted as we have argued in the last section. Thus it does not seem, at least in principle, possible to establish an ideal translation between any two theories.

As we have seen, the fundamental characteristic of a translation is to transfer existential requirements. That is, it should occur at the level of

---

[4]The article In defense of a dogma (Grice & Strawson, 1956) emphasizes this aspect criticizing Quine's TD.

the reference arrows, rather than occur at the level of the formulas. At the level of the formulas we hope to formalize an apparatus capable of fixing the desired transfer to the level of the reference arrows. If all relations of reference to objects of one theory can be converted into relations of reference to objects of another theory, we could say that the second manages to capture every ontological import of the other theory and thus manages to preserve the existential requirements of the first theory. We call this transfer of existential requirements: *ideal translation*. However, dealing directly with the idea of translation is to return to the same problem of fixing the intended model for a theory, since the relation of reference presupposes the two points of the arrow relation fixed: formulas and objects of the model.

2.1 Conditions for idealization

What is traditionally called translation is a mapping Map between formulas of a theory T_1 into formulas of another theory T_2 which (i) preserves some properties of the translated theory and, mainly, (ii) guarantees the result of relative consistency between the theories (if T_2 is consistent, then T_1 is consistent). We consider, in line with tradition, that a proof of relative consistency is strongly linked to the concept of translation. However, if we want to talk about the idea of *ideal translation*, we must minimize our expectations by stating that a mapping implies a (partial or total) aspect of the *ideal translation*.

This way of dealing with a concept puts us back on the floor and allows us to speak significantly of concepts such as translation, analyticity and meaning. To recover these concepts as idealizations is, at the same time, to recover the normative use of the dichotomy that they impose and to preserve the idea that we cannot perfectly fulfill their requirements. It is to recognize, then, that I can admit that there are no analytic sentences and at the same time say that philosophical and scientific activity are fundamentally different, because of the normative dichotomy imposed by the analytization or not of concepts. Indeed, there would be no fundamental difference between any sentence of a philosopher and of a scientist – all being subject to empirical revision – but there would be a fundamental difference in attitude to each sentence. On the concept of "bachelor", one prefers to argue about the sentence "bachelors are generally younger than married men" and the other prefers to discuss whether "bachelors are unmarried" – and we all know who is one and who is the other.

On What Counts as a Translation

It is not, though, allowed to any concept $\mathcal{C}$ the possibility of idealization. It is necessary (1) that there may be good reasons to say that certain X's are more $\mathcal{C}$ than other Y's; and (2) that the context in which the comparison is done is subject to less relativity than the context of the statement of $\mathcal{C}$. In fact, a large portion of philosophical concepts can be justified precisely in this way, and the idea of inscrutability of meaning and analyticity is simply the result of the nonobservance that ultimately any philosophical concept would be subject to a greater or smaller degree of inscrutability. If there are no good reasons for a α or β sentence to be ideally analytical, there are good reasons to say that $\alpha \vee \beta$ is more analytic than α and also than β; if there is no good reason to say that the extensional meaning of a sentence is measured by Tarski's semantics, there are good reasons to say that other options are bad; if, finally, there are good reasons to say that interpretations fail to capture everything a translation should accomplish, there is good reason to say that interpretations do this better than the simple mapping of true sentences into true sentences.[5]

It is from this argument that we recover the translation: although there is no mapping between formulas of a language in formulas of another language that completely transfers the existential requirements of a theory in the first language to the existential requirements of a theory in the other language, there are still reasons to understand certain methods as capturing aspects of translation neglected by other methods. This, on the other hand, makes us liberalize what counts as "translation". Any mapping that in some sense transfers ontological commitments between two theories can count as a translation. The point here is the observation that certain reductions do not count as a complete transport from one ontology to another, and a more accurate analysis on how this imperfection contaminates the analysis of ontologies is necessary. If, however, we suppose that interpretations are the only legitimate method of translation, then the case in which neither of the two theories interprets the other would be intractable; in this case, the liberalization of what counts as a translation can offer ways to compare ontologies where it was not possible.

---

[5]Benacerraf (1965) stresses a view, related to this argument, that I find inspiring. For him, although numbers are not sets, it does not mean that taking them as such is worthless: this strategy "cast some sobering light on what it is to be an individual number".

3 The sense of translation in a proof of relative consistency

What should we preserve in the mapping between the formulas of two theories so that it is possible to affirm that there is a transfer of existential requirements? One possible answer to this question is simply to state that all the true formulas of a theory must be mapped into true formulas of another theory. This results from the observation that the opposite seems absurd: if "$2 + 2 = 4$" is taken to be a false sentence of any theory, then no translation occured. However, the very notion 'true formulas' is subject to indetermination. Do we take as "true" all the formulas that are true in some model? We take as "true" the formulas that are true in an intended model? We take as "true" only formulas that are theorems of a theory? It is at least reasonable to admit that "$2 + 2 = 4$" is translated into a formula that one knows is not false. It is also acceptable to admit that "$2 + 2 = 4$" is translated into a true formula in the "intended model", as it is to be translated into a theorem.

We suppose, provisionally, that all theorems of a T_1 must be mapped in theorems of a T_2. As well noted by the tradition in the subject, a mapping that imposes only this restriction is too flexible: suffice, for example, that all the theorems of T_1 in some theorem of T_2 have been mapped. However, by imposing the condition that if α is mapped to β, then $\neg\alpha$ is mapped to $\neg\beta$ - we are able to perform some ontological analysis. The case where T_1 is inconsistent no longer supports such a mapping in a consistent theory T_2. Notably, this means that an inconsistent theory is ontologically irreducible to a consistent theory. In fact, an inconsistent theory commits itself to the existence of any object, whereas a consistent theory is only committed to the existence of a particular universe of objects.

Other types of requirements can be imposed for mappings so that we can affirm closer links between ontologies. We can impose the preservation of the boolean structure, as we can impose that existential quantifications are maintained existential quantifications. Each of these constraints has a role in effecting the transfer of ontological commitment, and in general we will say that the sum of all these constraints results in the mapping method.

In particular, a mapping method frequently used in the study of ontological reductions: relative consistency proofs (RCP). By the first order completeness theorem, we know that the consistency of a theory implies an ontology. For this reason, the proof that the consistency of a theory T_1 implies the consistency of a theory T_2 is a good reason to assume that the ontology of T_2 is reducible to the ontology of T_1. However, as we saw in the previous sections, this is not enough to affirm the reduction. In this case, we

On What Counts as a Translation

will call the *the sense of translation* of a RCP what counts as transference of ontological commitment in this RCP.

3.1 The general scheme for translation

Interpretations impose excessive restrictions on what counts as a translation. This method requires that each α formula of a T_1 theory be interpreted as a unique formula α^I in T_2's language. The procedure for determining α^I is

1. Regular: predicates, constants, and functions of T_1 are interpreted by predicates, constants, and functions definable in T_2.

2. Uniform: predicates, constants, and functions are always interpreted in the same way, regardless of where or how they occur in the T_1 formulas.

3. Universally regular: interpreted quantifiers are quantifiers limited by a single predicate in T_2.

In the treatment of natural languages, we are accustomed to make translations in which the context weighs substantially on the process that generates the translated sentence. Indeed, the case where the exact words of the dictionary substituted in one sentence form a sentence in the other language with the desired meaning is of a special type and is usually associated with rather simple constructions of both languages. However, in formal languages the requirement for uniformity seems more natural, though not necessary: it is not at all strange to suppose a translation of the relation of membership which means something when we speak of "sets of one kind" and another when we speak of "sets of another kind".

A Similar argument can be used to deny the necessity of universal regularity. It is possible to imagine that the context of the quantifiers changes according to the sentence that one wishes to translate. The predicate that defines the universe of interpretation could be variable according to the formulas being analyzed. Of course, it is necessary that this universe alternates in an ordered way, maintaining the necessary cohesion so that the RCP is possibly obtained. However, nothing in principle prevents the universe of interpretation from varying with context.

Therefore, to create a flexible notion of interpretation, we start with the analysis of two antagonistic forces: make the conditions (2) and (3) more flexible, while maintaining a version of the theorem of interpretation in such a way that it still implies relative consistency.

The interpretation theorem states that if all the axioms of a theory T_1 are interpreted in theorems of a theory T_2, then all the theorems of T_1 are interpreted in the theorems of T_2. This means that the interpretation preserves the logical structure of the arguments in T_1. Similarly, we expect a flexible version to satisfy: If T_2 **sees as true** each axiom of T_1 **brought by the mapping into a universe comprehensible to** T_2, then the same holds for any theorem of T_1.

We convert this condition into symbolic language using the following notation: $\alpha^{Tr(T_2)}$ denotes "α brought into the comprehensible universe of T_2"; and $\vdash^s$ denotes "seeing as true" in some compatible fashion to the definition of $Tr(T_2)$.

It follows, therefore, in a symbolic language the general scheme of interpretation: given two theories T_1 and T_2 and being that, for every axiom α_i of T_1, $T_2 \vdash^s \alpha_i^{Tr(T_2)}$, then $T_1 \vdash \alpha \Rightarrow T_2 \vdash^s \alpha^{Tr(T_2)}$.

Finally, in order to obtain the RCP, it is enough to impose on the $\vdash^s$ and $Tr(T_2)$ the following condition: if $T_2 \vdash^s \alpha^{Tr(T_2)} \wedge \neg \alpha^{Tr(T_2)}$, then there is a formula β in T_2 such that $T_2 \vdash \beta \wedge \neg \beta$.

A method that satisfies these conditions presents a great claim for transferring existential commitments. However, it is still necessary to answer whether or not there is a method that satisfies those requirements other than the method of interpretation itself. In fact, it is possible in this scheme to capture the RCP in which the assumption of the existence of a model for a theory implies the existence of a model for another theory.[6] However, this demonstration goes beyond the scope of this paper and will be presented in an upcoming article.

4 Final remark

In this article, we have insisted on Quine's strategy in ORel to show that not only ontology is relative, but the ontological reduction itself is relative. Nonetheless, instead of denying the meaningfulness of the use of the expression "the translation", we take the concept as a normative idealization. This approach allows us to come up with a more comprehensive plurality of translation methods – each of them having some (always partial) sense of translation. Eventually, as a result of this view, we may achieve some

---

[6]Many of the relative consistency proofs by model-theory are reducible to proofs by interpretations. But this is not unrestricted. An example for this is Novak's proof of equiconsistency between NBG and ZFC.

ontological comparison where it was not possible. This picture, thus, comprises the understanding that more than one method can count as partially transferring the existential requirements.

References

Benacerraf, P. (1965). What numbers could not be. *The Philosophical Review*, 74(1), 47–73.
Boghossian, P. A. (1996). Analyticity reconsidered. *Noûs*, 30(3), 360–391.
Freire, R. A. (2017). *Interpretation and truth in set theory.* (preprint)
Grice, H. P., & Strawson, P. F. (1956). In defense of a dogma. *Philosophical Review*, 65(2), 141–158.
Hamkins, J. D., & Yang, R. (2014). Satisfaction is not absolute. *to appear in the Review of Symbolic Logic*, 1–34. Retrieved from http://jdh.hamkins.org/satisfaction-is-not-absolute
Putnam, H. (1962). The analytic and the synthetic. In *Putnam 1975* (p. 215-227).
Quine, W. V. (1948). On what there is. *The Review of Metaphysics*, 2(1), 21–38.
Quine, W. V. (1968). Ontological relativity. *the Journal of Philosophy*, 65(7), 185–212.
Quine, W. V. (2000). Two dogmas of empiricism. *Perspectives in the Philosophy of Language*, 189–210.
Quine, W. V., Churchland, P. S., & Føllesdal, D. (2013). *Word and Object*. Cambridge: MIT press.
Shoenfield, J. R. (1967). *Mathematical Logic* (Vol. 21). Addison-Wesley Reading.

Alfredo Roque Freire
State University of Campinas
Institute of Philosophy
Brazil
E-mail: alfrfreire@gmail.com

Extensions and Projections in Deontic Default Logic

ANDRÉ FUHRMANN

Abstract: It will be argued that John Horty's proposal for deontic default logics does not extend beyond very simple default theories without losing its intended interpretation. The principal impediment can be removed by basing default inference on projections rather than extensions.

Keywords: Default logic, Deontic logic, Imperatives, Obligations, Conditional obligations, Default priorities

1 Introduction

This is a brief report summarising some findings concerning John Horty's proposal for deontic default logics (see Horty, 2012, 2014, and the bibliography therein). It will be argued that Horty's proposal does not extend beyond very simple default theories without jeopardising the intended deontic interpretation. The principal problem is what will be described in Section 3 as the Reflexivity Problem. For a more detailed exposition of the problem and possible responses see Fuhrmann (2017). In Section 4 I shall make a proposal so as to bypass the Reflexivity Problem. The solution consists in substituting the usual inference relation of Default Logic, defined in terms of extensions, by an inference relation based on quantification over a slightly different family of sets (projections). We shall see that the Reflexivity Problem can indeed not arise and, in Section 5, that the intended interpretation can be maintained once we move to more complex default theories, including those that contain information as to the order in which defaults should be considered.

2 Background

Let FML be the set of formulae of some propositional language containing a complete set of Boolean connectives. A (simple) default relation D is a subset of FML^2. Where a and b are formulae, we also write $a \Rightarrow b$ for $(a, b) \in D$ and call $a \Rightarrow b$ a *default*. A *default theory*, in the sense of Reiter (1980) and others, is a pair (D, A), where D is a default relation, i.e. a set of defaults, and A is a subset of FML, a set of *assumptions*.

Given a default theory (D, A), its set of assumptions A can be extended by "drawing on" defaults in D. The idea is to look for defaults whose premiss is in A and then include the conclusion in the extended set of assumptions A', if the result is consistent. Then use A' to detach further default-conclusions, if this can consistently be done. Continue the process until the defaults have saturated the assumptions, i.e. until the set of assumptions can no longer be consistently extended in the above manner. A set of formulae that maximally extends A with the aid of D is called an *extension* of (D, A).

In Section 4 we shall briefly recall a precise definition of the notion of an extension. Here it suffices to take note of two immediate and well-known consequences of the basic idea as sketched above. First, the result of the extension process may depend on the order in which defaults are considered for the purpose of detachment. To illustrate with a small example, let $(D = \{a \Rightarrow b, a \Rightarrow \neg b\}, A = \{a\})$. Now start with $a \Rightarrow b$. Then b will be detached and the second default cannot consistently be used. If we start with $a \Rightarrow \neg b$ instead, then $\neg b$ will be detached and the first default cannot be used. So two different and mutually inconsistent extensions will be produced, depending on the order in which the defaults are used. Given a default theory, its set of extensions $Ext(D, A)$ is in general not a singleton set. The set $Ext(D, A)$ can be reduced by forcing the order in which defaults are to be activated. In this case we think of D as structured by some ordering $<$, thus working with *prioritised default theories*, usually represented by triples $(D, A, <)$.

Second, an extension E of a default theory (D, A) is an extension of its assumption set A. So we have quite trivially

Inclusion $\qquad\qquad\qquad A \subseteq E.$

Even only with the rough sketch above at hand we can say a bit more about extensions. Clearly, they are bound below by A. In the lucky case that all defaults can be used for the purpose of extending A, an extension will conjoin A with the set $Concl(D) = \{b : a \Rightarrow b \in D\}$ of all conclusions of D. Given that we wish to extract as much information as possible from A together with D, the upper limit of an extension must be $\mathrm{Cn}(A \cup Concl(D))$, the closure of $A \cup Concl(D)$ under logical consequence. In general then,

we invariably have
$$E = \text{Cn}(A \cup Concl(D')),$$
for some $D' \subseteq D$. (Variation comes in by choosing constraints that determine D'.)

What can we *infer* from a set of assumptions A given a set D of defaults? *Default Logic* (DL) proposes to look at answers in which we quantify over $Ext(D, A)$. Two answers are salient:

(i) $A \mathrel{\mid\!\sim}_D^d x$ iff x is in *every* extension of (D, A) — skeptical, cautious, disjunctive mode of inference (the definition usually found in the DL-literature);

(ii) $A \mathrel{\mid\!\sim}_D^c x$ iff x is in *some* extension of (D, A) — credulous, brave, conflict mode of inference.

Note that in virtue of Inclusion both these "modes" of default inference satisfy

Reflexivity $\qquad\qquad A \mathrel{\mid\!\sim}_D x, \ \forall x \in A.$

Unless noted otherwise, we shall focus on default inference in the sense of (i). We may therefore drop the superscript d and think of $\mathrel{\mid\!\sim}$ as standing for default inference in the disjunctive mode.

3 Horty's default evaluation rule

Where a is formula, we may represent the (unconditional) imperative "See to it that a be the case!" by $!a$. Let I be a set of such imperatives. Now suppose that we generate a set of defaults from I such that

$$\top \Rightarrow x \in D(I) \text{ iff } !x \in I$$

There is a straightforward sense in which, according to I, it *ought* to be that x just in case $\emptyset \mathrel{\mid\!\sim}_{D(I)} x$. We may also say: the default theory $(D(I), \emptyset)$ *supports* the assertion that it ought to be that a just in case a can be derived ($\mathrel{\mid\!\sim}$) by using the defaults in $D(I)$. Thus, if $D = D(I)$ and $A = \emptyset$, then we may give the predicate ϕ a deontic interpretation in the following biconditional:

(DER) $\qquad\qquad (D, A)$ supports ϕx iff $A \mathrel{\mid\!\sim}_D x.$

André Fuhrmann

This is, in effect, the *default evaluation rule* for deontic modals used in (Horty, 2012).[1] It transports into the framework of DL an idea that goes back to van Fraassen (1972, 1973).

If we apply (DER) to default theories in which all defaults are of the form $\top \Rightarrow x$ and in which there are no assumptions, then the extensions will be the maximally consistent subsets of $Concl(D)$. To briefly return to the two modes of inference, if there are conflicting extensions, say one containing a, the other containing $\neg a$, then in the disjunctive (skeptical) mode, neither a nor $\neg a$ enjoys the property ϕ. If, on the other hand, we use the conflict (credulous) mode, then we have conflicting oughts, i.e. both ϕa and $\phi \neg a$.

In Horty's work such simple default theories are designed as mere stepping stones. They illustrate the basic idea to be preserved when considering more interesting default theories. Thus, Horty (2014, p. 438) writes:

> These [simple imperative default theories] are very simple, of course, but the normative interpretation can be generalised to richer theories as well—theories of the form $(D, A, <)$ in which the hard information from A may not be empty, the defaults from D might have nontrivial premises, and there might be real priority relations among them.

It is difficult, however, to see how this generalisation, preserving the initial idea, should be possible. The difficulties are detailed in Fuhrmann (2017). The principal stumbling block is as follows.

Consider the simplest case of a default theory with non-trivial default premises and a non-empty assumption set: $(D = \{a \Rightarrow b\}, A = \{a\})$. By Reflexivity, we have $A \mathrel{\vert\!\sim}_D a$, whence ϕa by the evaluation rule (DER). But a is a fact-stating assumption, or so we may suppose. So ϕ cannot carry a deontic interpretation—unless we declare factual assumptions to be obligatory. The argument can be sidestepped by restricting A to assumptions generated from imperatives. But this is not really an option, for defaults could then only be triggered if they are generated from conditional imperatives in which the condition happens to be commanded. Though examples of such coincidences can be made up, the approach would be of little interest if it *required* that conditions and commands always coincide in this way. Call this train of thought the *Reflexivity Problem*.

---

[1] Horty formulates the rule differently, using $\mathrel{\vert\!\sim}$ for the relation of support (and $\bigcirc$ for ϕ). But since support is defined in term of default inference, and since we have reserved $\mathrel{\vert\!\sim}$ for the latter, we better not overload $\mathrel{\vert\!\sim}$ by using it also for the former. Note also that without further ado and in contrast to the ought-operator in modal deontic logic the predicate ϕ is not iterable.

The problem is aggravated when we try to implement priorities among defaults in a more flexible way than the one briefly mentioned above. Instead of fixedly structuring D by an order $<$, we can incorporate priority information in the assumption set and let such priorities occur as premisses or conclusions of defaults. In this way priorities can themselves be inferred by default inference. This requires an extension of the language by ordering propositions $d \prec d'$, where d and d' name defaults. Since all ordering propositions now occur in A, they enjoy deontic status (by (DER) & Reflexivity). But the ordering of defaults is just meant to help determining one's obligations—the ordering is not itself obligatory, not in general anyway.

4 Projections versus extensions

If we are to generalise Horty's approach beyond toy default theories of merely heuristic value, we need to solve the Reflexivity Problem. For, in the presence of Reflexivity for $\mathrel{\mid\!\sim}$ the evaluation rule (DER) overgenerates obligations. So we need to filter out the overgenerated items.

A first, simple approach to solve the problem is to supplement the right-hand-side of (DER) by a clause that aims at taking out the overgenerated items:

(†) $(D(I), A)$ supports Ox iff $A \mathrel{\mid\!\sim}_{D(I)} x$ *unless* $A \vdash x$.

Now we can populate A with factual assumptions without having Oa for all $A \vdash a$. We can also use these assumptions to trigger real defaults $a \Rightarrow b$ as generated from conditional imperatives of the form $!b/a$. Categorical obligations are encoded as before by defaults of the form $\top \Rightarrow b$. Moreover, since $A \vdash \top$ (for any A), we have the welcome side-effect that logical truths are never obligatory. (By contrast, the somewhat strange $O\top$ cannot be avoided in standard deontic logic where O is treated as a normal modal operator.)

Since we work under the assumption that defaults are the only deontically loaded items in the theory (D, A), nothing that can be inferred without their participation can enjoy deontic status. The unless-clause in (†) has the effect of filtering out propositions that can be inferred without triggering defaults. So we may expect that this clause removes the undesired items. Although this much is true, (†) overshoots the mark as the following example shows.

The default theory ($D = \{a \Rightarrow b\}$, $A = \{a \rightarrow b, a\}$) has only one extension $E = \text{Cn}(a \rightarrow b, a)$. On the one hand, since b can be inferred from A alone, Ob is not supported according to (†). On the other hand, given that there is a bijection between the underlying set I of imperatives and the defaults in D, the presence of $a \Rightarrow b$ in D implies that there is a conditional imperative $!b/a$ in I. Since the condition a obtains, the default $a \Rightarrow b$ can be triggered, giving b. So we should expect that Ob is supported by the theory—contrary to what (†) rules. We have thus found that (†) undergenerates! We need a better way of solving the Reflexivity Problem.

The idea of the following proposal is simple. Defaults $a \Rightarrow b$ represent conditional imperatives. So only the conclusions of defaults should fall into the scope of the derivative ought-predicate, and no conclusion of a triggered default should be left out. Thus, we are looking for a function that partially projects a set of defaults, i.e. of pairs (x, y), to their right-hand elements, y. The input to such a projection is, apart from D, the set A of assumptions which trigger defaults and an ordering in which the defaults are to be considered for triggering. We here implement this idea by adjusting the inductive definition of an extension in DL as first proposed by Brewka (1994) (see also Makinson, 2005).

Let (D, A) be a default theory and let $(D, <)$ be a strict total order of the defaults. Since we assume D to be countable, we may think of such ordering as an indexing of the defaults by the natural numbers in their natural sequence. We start the construction of the projection of D by A under $<$ by putting

$$A_0 = \emptyset.$$

(In the definition of extensions we would start with $A_0 = \text{Cn}(A)$ instead.)

In the step A_{k+1} we look for the first default $x \Rightarrow y$ in $(D, <)$ such that (i) $y \notin A_k$, (ii) $A \vdash x$, and (iii) $A_k \nvdash \neg y$. If there is such a default, then we put

$$A_{k+1} = \text{Cn}(A_k \cup \{y\});$$

otherwise we let $A_{k+1} = A_k$, thereby ending the construction. (In the definition of extensions we would replace (ii) $A \vdash x$ by (ii) $x \in A_k$.)

Finally we sum up:

$$A_{(D,<)} = \bigcup \{A_i : 0 \leq i \leq \omega\}.$$

P is a *projection* of D by A iff $P = \text{Cn}(A_{(D,<)})$, for some default ordering $(D, <)$.

Extensions and Projections

The process of constructing a projection is just as cumulative as is the construction of extensions, i.e. in both cases we have

$$A_0 \subseteq A_1 \subseteq \cdots .$$

But unlike in the case of extensions, default-conclusions detached at one stage in the construction cannot be used as premises to trigger defaults in later stages. This is as it should be, since default-conclusions represent imperatives, not factual assumptions that could be used to match the hypothesis of a hypothetical imperative. Factual assumptions reside only in the set A, as reflected above in the condition (ii).

About *extensions* of A by D recall that

$$E = \mathrm{Cn}(A \cup Concl(D')),$$

for some $D' \subseteq D$ and $Concl(D') = \{b : a \Rightarrow b \in D'\}$. Thus, $A \subseteq E$, which generates the Reflexivity Problem. About *projections* of D by A note that

$$P = \mathrm{Cn}(Concl(D')), \text{ some } D' \subseteq D.$$

Thus, typically we do not have $A \subseteq P$. We now replace extensions by projections in the definition of $\mathord{\sim}$ (in both modes) and thus define a new pair of relations $\mathord{\sim}^*$ as follows:

(i) $A \mathrel{\vert\!\sim}_D^{*d} x$ iff x is in *every* projection of (D, A) — the disjunctive mode;

(ii) $A \mathrel{\vert\!\sim}_D^{*c} x$ iff x is in *some* projection of (D, A) — the conflict mode of inference.

Finally, we propose a new evaluation rule (based on a set I of hypothetical imperatives) for the ought-predicate O:

(DER*) $\qquad (D(I), A)$ supports Ox iff $A \mathrel{\vert\!\sim}_{D(I)}^{*} x$.

Since the ought-predicate is now determined by quantifying over projections, we know that (D, A) supports Ox only if $x \in \mathrm{Cn}(Concl(D'))$ for some $D' \subseteq D$. Thus overgeneration cannot arise: mere assumptions cannot gain ought-status. Assumptions can gain such status only if they also occur as conclusions of defaults—which is as it should be.

How does (DER*) treat the undergeneration example above? In the example we have $D = \{a \Rightarrow b\}$, $A = \{a \to b, a\}$ and (†) does not deliver Ob. Since there is only one default, there is only one (trivial) ordering to consider.

André Fuhrmann

$A_0 = \emptyset$.
$A_1 = \{b\}$, since $a \Rightarrow b \in D$, (i) $b \notin A_0$, (ii) $A \vdash a$, and (iii) $A_0 \not\vdash \neg b$.
$A_2 = A_1$, since no defaults apply.

Thus $P = \mathrm{Cn}(b)$ is the only projection, whence $A \mathrel{\vert\!\sim}_D^* b$ (in both modes of inference). That is to say, Ob is supported by (D, A) while Oa is not. (We note in passing that $O\top$ is also supported, by the logical closure of projections. So we lose the welcome side-effect of the otherwise less felicitous rule (†).)

Projections are the result of using assumptions so as to detach the conclusions of default rules—to project the rule (x, y) to y. The interpretation of projections naturally depends on what we assume about the interpretation of defaults. Above we have followed Horty's idea—inspired by van Fraassen—that defaults are intimately related to conditional imperatives: that there is bijection between the two. But this is certainly not the dominant interpretation—better, perhaps: heuristic—considered in the literature. According to the standard interpretation of DL, defaults represent risky inference tickets. These are licences to proceed from premises to conclusions with a *caveat*. Consequently, conclusions only reached by using such rules inherit the vulnerability of the rules used. Assumptions, by contrast, are treated as safe by hypothesis. Under this interpretation, a projection of A by D represents the *risky* information that can be extracted from (D, A) given a fixed ordering of the defaults; an extension, by contrast, represents the *total* information, risky or safe, implicit in (D, A) relative to an ordering of D. Once the effects of particular orderings are cancelled out by quantification, we arrive at default inference in terms of projections. Under the interpretation at hand, such inference represents strictly risky inference. As far as one sometimes wishes to know whether information extracted from a default theory is safe from or vulnerable to an increase of assumptions, such a notion of inference in terms of projections can be useful.

We can sense a general idea at work in the last paragraph. Defaults can be seen as transforming into propositional rules certain conditional speech acts. (Recall that these need not be substantially conditional: the condition can be vacuous, i.e. $\top$.) If the defaults are then applied to assumptions, we derive propositions which fall under a predicate that is obtained from the character of the acts considered. In the standard interpretation of DL, the act is that of assertion and the predicate is truth. In this case the evaluation rule (DER) is of little interest. For given that truth is redundant in the sense

Extensions and Projections

of the biconditional $true(x)$ iff x, the rule just comes to this:

$$(D, A) \text{ supports } x \text{ iff } A \mathrel{\mid\!\sim}_D x.$$

Things are different if we base D on a set of conditional commands. Suppose we treat commands in the Fregean manner, i.e. as applying a particular commanding "force" to a propositional content (Frege in "Der Gedanke", 1918). We can isolate the propositional content and transform conditional commands into default rules which, in turn, can be applied to assumptions. A judiciously chosen evaluation rule—the proposal offered here is (DER*)—can then reveal those propositions that defeasibly enjoy ought-status on the basis of the commands issued and the assumptions made. The approach can be seen as solving, by brute regimentation, the Frege-Geach problem; cf. Geach (1960, 1965). The same recipe can be applied to other Fregean forces: wishes uttered, questions raised, damnations expressed, and so on. Here is an incomplete table of correspondences employable in this manner:

Force	Act	Predicate ϕ
Assertive	Jack asserts that p	p: true
Imperative	Jack commands that p	p: ought to be
Optative	Jack wishes that p	p: wished for
Interrogative	Jack asks whether p	p: asked whether
Damnative	Jacks boohs p	p: damned

5 Conditional obligations and ordering defaults

5.1 Conditional obligations

Once the basic repair is done as above we can proceed to consider Horty's evaluation rule for conditional obligations – but now with $\mathrel{\mid\!\sim}^*$ replacing $\mathrel{\mid\!\sim}$ in the original version:

(C*) $\qquad (D, A) \text{ supports } O_y x \text{ iff } A, y \mathrel{\mid\!\sim}_D^* x$

The original version overgenerated conditional oughts by the Reflexivity Problem: mere factual assumption automatically gained ought-status under arbitrary conditions. Thus, in particular, if $x, y \in A$, then we have both $O_y x$ and $O_x y$. This is now prevented by working with projections rather than extensions.

André Fuhrmann

5.2 Ordering defaults

We here consider only the more interesting case of extending the language by priority propositions (PPs) of the form $d \prec d'$, where d and d' are names of defaults in D and $\prec$ is a predicate applying to pairs of such names. For details as to how PPs are employed in DL see e.g. Makinson (2005). The basic idea is as follows. Let (D, A) be a default theory possibly containing PPs as constituent formulae in D or A. If we ignore what the PPs express, then the set *Ext(D,A)* of extensions is determined by all default orderings $(D, <)$. If we take heed of the PPs, then certain orderings should be disconsidered. For example, suppose that we derive $d \prec d'$ from A, possibly using D, where d and d' name the defaults δ and δ' respectively. Then an extension in which we apply δ' before δ would disrespect the information $d \prec d'$, whence, it is not admissible. The inference relation $\mid\!\sim$ should thus be defined in terms of admissible extensions.

Let us now consider the three places where PPs can occur: as assumptions; as premisses of defaults; as conclusions of defaults. First, assumptions. The evaluation rule (DER) in terms of extensions, suffering from the Reflexivity Problem, would give all PPs deontic status: they all ought to be the case. But this seems wrong. PPs should help us to determine what our obligations are, they are not in general themselves obligatory. The imperatives to which they can be taken to relate – "consider δ_1 before δ_2!" – are typically quite different from the imperatives we are interested in here. The latter are based on appraisals of acts; the former reflect the relative merits of imperatives of that latter kind. The evaluation rule (DER*) in terms of projections keeps PPs out of the scope of the O-predicate—as long as they do not themselves reflect commands (see below). On the other hand, (DER*) allows to let PPs do the work they are designed to do in a way that is consonant with the deontic interpretation under which we here consider default theories. To take a very simple example, in the default theory

$$D = \{\top \Rightarrow a, \top \Rightarrow \neg a\}, \ A = \{(\top \Rightarrow a) \prec (\top \Rightarrow \neg a)\}$$

we have encoded two conflicting commands together with information, in form of a PP, as to how the conflict should be resolved. Given the PP in A, the theory has only one admissible extension, viz. $\mathrm{Cn}(a)$, whence Oa rather than $O\neg a$ is supported in both the conflict and the disjunctive mode of inference ($\mid\!\sim^*$).

Next, PPs as premisses of defaults. These are of the form $(d_1 \prec d_2) \Rightarrow a$. We are assuming here that each default reflects a hypothetical imperative.

We therefore need to drive the bijection between defaults and imperatives into the premises of defaults if these happen to be PPs. Thus the default $(d_1 \prec d_2) \Rightarrow a$ corresponds to an imperative $!a/i_1 \prec i_1$ where i_1 is itself of the form $!b/c$ (and likewise i_2). These are complicated imperative phrases but they do not sound confused: see to it that a given that you prefer the one command over the other.

Finally, PPs as conclusions of defaults, i.e. defaults of the form $a \Rightarrow (d_1 \prec d_2)$, reflecting an imperative $!(i_1 \prec i_2)/a$ (with i_1 and i_2 further imperatives). The content of the command is an act of preference or choice, as in "better take the train than the car!" ("given that you are late"). So here we have a case where a default theory can support the subsumption of a PP under the ought-predicate.

6 Conclusion

I have argued that Horty's project of a Deontic DL gets stuck right after the start. The principal problem is the inclusion property of extensions: Extensions contain the assumptions they extend. But these assumption typically do not have deontic status—they are no oughts. This is the Reflexivity Problem. The problem can be solved, if we move from extensions (of A by D) to projections (of D by A). The solution continues to support the intended interpretation if we move to considering an evaluation rule for conditional obligation or to default theories that include priority information.

References

Brewka, G. (1994). Reasoning about priorities in default logic. In *Twelfth National Conference on Articial Intelligence* (pp. 940–945).

van Fraassen, B. C. (1972). The logic of conditional obligation. *Journal of Philosophical Logic, 1*, 417–438.

van Fraassen, B. C. (1973). Values and the heart's command. *Journal of Philosophy, 70*, 5–19.

Fuhrmann, A. (2017). Deontic modals: Why abandon the default approach. *Erkenntnis, 82*, 1351–1365.

Geach, P. (1960). Ascriptivism. *Philosophical Review, 69*, 221–225.

Geach, P. (1965). Assertion. *Philosophical Review, 74*, 449–465.

Ginsberg, M. L. (1987). *Readings in Nonmonotonic Reasoning*. San Francisco: Morgan Kaufmann Publisher Inc.

André Fuhrmann

Horty, J. F. (2012). *Reasons as Defaults*. New York: Oxford University Press.
Horty, J. F. (2014). Why abandon the classical semantics? *Pacific Philosophical Quarterly*, 95, 424–460.
Makinson, D. (2005). *Bridges from Classical to Nonmonotonic Logic*. London: King's College Publications.
Reiter, R. (1980). A logic for default reasoning. *Artificial Intelligence*, 13, 81–132. (Reprinted in Ginsberg, 1987, pp. 68–93)

André Fuhrmann
Goethe-Universität
Institut für Philosophie
Germany
E-mail: `fuhrmann@em.uni-frankfurt.de`

What Makes True Universal Statements True?

BOB HALE[1]

1 Outline

What makes true universal statements true? In discussing this question, I shall be especially interested in how it is to be answered within the framework of what Kit Fine calls *exact truth-maker semantics*.[2] In an extremely useful survey article Fine (2017) locates exact truth-making within the more general truth-conditional approach to semantics as follows. First, we distinguish between *clausal* approaches such as Davidson (1967), on which truth-conditions are not given as non-linguistic entities but by clauses through which a theory of truth specifies when a statement is true, and *objectual* approaches, according to which truth-conditions are not clauses but worldly entities which stand in a relation of truth-making to the statements they make true or false. Then, within objectual approaches, we may distinguish those which take truth-makers to be *possible worlds* (as in what is generally known as possible world semantics) from those which take them to be *states* or *situations*, 'fact-like entities that serve to make up a world, rather than worlds themselves' (Fine, 2017, p. 557), as in the situation semantics developed by Barwise and Perry (1983). Finally, we may distinguish,

---

[1] Thanks to Vladimir Svoboda and Vit Puncochar for inviting me to give a talk at the 2017 Logica symposium, and to those who contributed to its discussion, and especially to Ilaria Canavotto for pursuing some issues further in correspondence. I should like also to express my gratitude to participants in workshops on modality and truthmaking and on generality held in Augsburg and Oslo, at which I presented versions of this material, and especially to Kit Fine, Wes Holliday, Jon Litand, Øystein Linnebo, and Ian Rumfitt. I regret that limitations on length have prevented me from responding here to the challenging points and questions they raised. I hope to address them in a fuller discussion of the approach for which this paper argues.

[2] In thinking about this topic, I have benefited enormously, not only from reading Kit Fine's recent papers (Fine, 2017 and Fine, in press), but from some very useful discussion with him. My understanding of the truth-maker approach to semantics derives almost entirely from the clear, straightforward, and searching introduction to it that his work provides. Indeed, it was reading Fine (in press) which made me aware of the interest and importance of the the leading question of this paper, and first got me thinking seriously about it.

within 'stately' (as opposed to 'worldly') semantics, between *loose, inexact,* and *exact* forms of truth-maker semantics. *Loose* truth-making is a purely modal relation: a state s makes true, or verifies, a statement p just in case it is impossible that s should obtain without p being true. By contrast, both inexact and exact verification require there to be a relevant connection between the state s and the statement p: s is an *inexact* verifier for p if and only if s is at least partially relevant to p, and is an exact verifier if and only if it is *wholly* relevant.[3] Any exact verifier is also an inexact verifier, and any inexact verifier is also a loose verifier, but the converses do not, of course, hold.

On the version of exact truth-making which Fine presents in this and other recent papers, a true universally quantified statement, $\forall x B(x)$, is made true by the verifiers of its instances – or a little more precisely, it is made true by a state which is composed of the states which verify its instances. Essentially the same account is endorsed by nearly all other advocates of truth-maker semantics whose work I have consulted, so I think it may fairly be labelled the standard account. Its widespead acceptance is hardly surprising, since it can appear both very natural and even inevitable, given that $\forall x B(x)$ is true iff $B(x)$ is true of every object which is an admissible value of its free variable. But whilst I think there is a subclass of universal statements for which the standard account is correct, I do not think it gives the right account of truth-making for universal statements in general.

My main aim here is to promote an alternative account of the truth-makers for quantified propositions. I shall also give some attention to two closely related questions: first, when, and why, we should favour an alternative to the standard account, and second, whether the alternative account I favour can be accommodated within the framework of *exact* truth-maker semantics, in Fine's sense.

What follows divides into two parts. In the first, I make a case for an alternative account of what makes some universal statements true, prescinding altogether from the questions how, in detail, that account might run, and whether and, if so, how it may be implemented within the framework of exact truth-maker semantics. In the second, I turn to those questions.

---

[3] What, more precisely, is to be understood by this will be discussed in due course.

What Makes True Universal Statements True?

2 A case for an alternative account

2.1 The standard account and its shortcomings

As Fine observes, a major difference between possible world semantics and state or situation semantics concerns completeness – worlds are standardly taken to be complete, or maximal, in the sense that they determine the truth-value – true, or false – of every statement of the language for which a semantics is being given, whereas states or situations settle the truth-values of particular statements, but typically leave those of many others unsettled.[4] One consequence of this difference is that in truth-making semantics we need to take account both of states which make statements true (their verifiers) and of states which make them false (their falsifiers). If a statement p is not true at a possible world w, it will be false at w; but a state s which does not verify p may not falsify it either, for it may – and typically will – simply leave its truth-value unsettled. Thus the clauses for conjunction, for example, will tell us that state s verifies $A \wedge B$ iff it is the fusion $s_1 \sqcup s_2$ of two states, s_1 and s_2, one of which verifies A and the other B; and that s falsifies $A \wedge B$ iff s either falsifies A or falsifies B. The usual treatment of universal quantification takes a state s to verify a universal quantification $\forall x A(x)$ iff s is the fusion $s_1 \sqcup s_2 \sqcup \ldots$ of states $s_1, s_2, \ldots$ which verify $A(t_1), A(t_2), \ldots$ where $A(t_1), A(t_2), \ldots$ are all the instances of $\forall x A(x)$, i.e. s verifies the conjunction $A(t_1) \wedge A(t_2) \wedge \ldots$ of all the instances; and s is taken to falsify $\forall x A(x)$ iff it falsifies one of $A(t_1), A(t_2), \ldots$.[5,6]

---

[4]This is equally true of alternatives to full-blooded world semantics which replace complete worlds by possibilities which are typically incomplete, such as the possibility semantics advocated in Hale (2013, ch. 10). For somewhat different developments of the same general approach, see Humberstone (1981), where the idea makes its first appearance, and more recently in Rumfitt (2015, ch. 6).

[5]Here we assume for simplicity that all the objects in the domain of the quantifier are denoted by terms of the language. We could dispense with this assumption by stipulating that, for each element d of the domain $\mathcal{D}$, the fusion $s_1 \sqcup s_2 \sqcup \ldots$ is to have a part s_d which verifies $A(x)$ when d is taken as the value of x.

[6]See Fine (2017, §7). Assuming the quantifier ranges over the individuals $a_1, a_2, \ldots$ denoted by the constants $\mathbf{a_1}, \mathbf{a_2}, \ldots$, Fine suggests that 'we might take the content of $\forall x \varphi x$ to be the same as the content of the conjunction $\varphi(\mathbf{a_1}) \wedge \varphi(\mathbf{a_2}) \wedge \ldots$' and gives the clauses:
a state verifies $\forall x \varphi x$ if it is the fusion of the verifiers of its instances $\varphi(\mathbf{a_1}) \wedge \varphi(\mathbf{a_2}) \wedge \ldots$
a state falsifies $\forall x \varphi x$ if it falsifies one of its instances.
More formally:
$s \Vdash \forall x \varphi(x)$ if there are states $s_1, s_2, \ldots$ with $s_1 \Vdash \varphi(\mathbf{a_1}), s_2 \Vdash \varphi(\mathbf{a_2}), \ldots$ and $s = s_1 \sqcup s_2 \sqcup \ldots$
$s \dashv \forall x \varphi(x)$ if $s \dashv \varphi(\mathbf{a})$ for some $\mathbf{a} \in A$ (where A is the domain of individuals).

Bob Hale

There are, certainly, universal statements for which this account is very plausible. Consider, for example, the statement I might make that *all my children live in England*. I have just three children, Thom, Charlie and Josh. It is very plausible that my statement is true iff the conjunction: *Thom lives in England* ∧ *Charlie lives in England* ∧ *Josh lives in England* is true, and that it is made true by the state $s = s_1 \sqcup s_2 \sqcup s_3$, where s_1, s_2, s_3 verify its conjuncts.[7] I say only that it is very plausible, rather than it is evidently true, because, of course, it might be objected that the truth of the conjunction suffices for that of the universal generalization only given the additional facts that Thom, Charlie and Josh are my children and that I have no others, and that, accordingly, what makes the universal generalization true is not just the state s by itself, but this state together with the state which makes it true that these and only these are my children. Proponents of truth-maker semantics may diverge over whether the truthmakers for universal statements must always include such totality facts. I neither need, nor wish, to try to settle that disagreement here. For simplicity, I shall not take totality facts to be parts of the truth-makers for universal statements (or the falsity-makers for existential statements), but it will be clear enough that including them would do nothing to allay the doubts about the adequacy of the standard account which I shall be raising – indeed, including them would, if anything, make matters worse for that account.

It would be very easy to multiply examples of universal statements like this one, for which the standard account, plus or minus a totality fact, seems clearly right. But it seems to me equally easy to give examples for which that account seems more or less obviously wrong. For the moment, I shall give just a couple of examples, which may at first appear somewhat special cases, although, as I hope will become clear as we proceed, they are in essential respects representative of a large class of universal statements to which the standard account does no justice.

Consider first the statement that *every natural number has another natural number as its immediate successor*. According to the standard account, what makes this true is the fusion $s_1 \sqcup s_2 \sqcup \ldots$ where s_1 verifies the statement that 0 *is a natural number and is immediately followed by* 1, *which*

---

[7]This treatment of the example assumes that my general statement is to be analysed as formed with a restricted quantifier, to be treated in accordance with the modification Fine (2017, §7) proposes. Without this assumption, applying his clause for unrestricted quantification given in the previous note would take the instances to be the truth-functional conditionals $\mathbf{a_i}$ *is a child of mine* → $\mathbf{a_i}$ *lives in England* (one for each element of the domain of individuals) and would take the verifier for the general statement to be the fusion of the states which verify them.

What Makes True Universal Statements True?

is another natural number, and s_2 verifies the statement that 1 *is a natural number and is immediately followed by* 2, *which is another natural number*, and so on for an additional infinity of states s_i each verifying the corresponding statement that s_{i-1} *is a natural number and is immediately followed by* s_i, *which is another natural number*. The implausibility of this account does not derive, primarily or mainly, from its need to postulate an infinitary state – at least, this is not why I find it implausible, although there may be some who would wish to object to it on this ground, and there is a closely related further difficulty which I will come to shortly. What, primarily, seems to me implausible in the standard account is that it pays no attention to what appears to be an obvious uniformity in the grounds for the truth of the instances. As far as that account goes, the separate instances might just as well be made true by entirely separate and unrelated states concerning individual natural numbers, much as the instances of the general statement about my children are made true by separate and independent states concerning their whereabouts. Of course, the standard account does not *entail* that the states which make the instances of a universal generalization true are unrelated – but the point is that it pays no attention to such connections, even when they seem plainly relevant to what makes both the generalization itself and its instances true.

The further difficulty to which I alluded previously comes into play if it is held that the verifier for a universal statement, on the standard account, must incorporate a totality fact. On this view, the verifier for a universal statement will comprise – be the fusion of – *two* states: the state $s = s_1 \sqcup s_2 \sqcup \ldots$, where $s_1, s_2, \ldots$ are as before, together with a state t which verifies a statement to the effect that $o_1, o_2, \ldots$ are all the objects of the kind over which the universal statement generalizes. In our example, t will be a verifier for the statement that $0, 1, \ldots$ *are all the natural numbers*. The obvious worry is that that way of stating the totality fact is just a fudge – it relies on our understanding the dots, not as mere 'dots of laziness' but as standing in for an infinite list which we cannot complete.[8]

---

[8] It perhaps bears emphasis that this difficulty is entirely distinct from any concern that may be felt over the fact that the statement of the totality fact is itself a further universal statement. If it is held that every true universal statement is verified by a complex state involving a suitable totality fact, there is an obvious threat of vicious regress. Clearly this concern does not depend upon the infinity of the domain of the universal quantifier – it arises equally if the quantifier ranges over a finite domain, whether it be closed (as in the example about my children) or open (as with, for example, the statement that *all aardvarks are mammals*, assuming that there will one day be no aardvarks).

Bob Hale

I do not claim that this difficulty amounts to a decisive objection, nor that the relevant totality fact is unstateable. A resolute defender of the standard account with totality facts might insist that the existence of infinitary totality facts is one thing, and their stateability is another. Such a stance requires a robust, but familiar kind of realism, of the same stripe as appears to be involved in thinking that Goldbach's Conjecture may be insusceptible of any general, finitely articulable proof, but is nonetheless rendered true by an infinity of facts of the form: $2n$ *is the sum of primes p_i and p_j*. Further, it might be claimed that, while the totality fact in our example cannot be stated by completing the list in the would-be statement that $0, 1, \ldots$ *are all the natural numbers*, it can be stated readily enough in another way, viz. *0 and its ancestral successors are all the natural numbers*. But note that this is hardly a response with which a defender of the standard account should feel comfortable. For the fact that 0 and its ancestral successors are all the natural numbers is just the fact that everything which is a natural number is either 0 or the immediate successor of a natural number – and in the presence of this totality fact, the fusion $s = s_1 \sqcup s_2 \sqcup \ldots$, of states which verify the instances of our universal statement is clearly redundant, since the totality fact is by itself enough to verify it.

Before I venture any suggestion about what might better be taken to verify our statement about the natural numbers, let me give just one more example. Consider the statement that *(all) aardvarks are insectivorous*. According to the standard account, what makes it true is the fusion $s = s_1 \sqcup s_2 \sqcup \ldots$, of states which verify its instances a_1 *is an aardvark and eats insects*, a_2 *is an aardvark and eats insects*, ... – one for each aardvark.

This seems to me pretty incredible. As with the previous example, a determined defender of the standard account could just stick to his guns. But doing so seems to me to require chewing on some pretty indigestible bullets. One important feature of the example – and in this it is typical of a great many others which might be given – is that, in contrast with our purely numerical example, no actual or potential infinity of instances need be involved; what is involved is a possibly – indeed, almost certainly – *finite but open-ended* domain of entities. The generalization covers not only presently existing aardvarks and the large but finite number which have since perished, but also those which are yet to be born. In consequence, the state which verifies the generalization must be the fusion of states, an indefinite number of which verify instances concerning aardvarks which do not yet exist, but will exist at some time in the future. Unless a defender of the standard account is ready to insist that the generalization is not yet true, and will not be true at any future date before Armageddon, he must accordingly maintain either that there now exist states which somehow verify statements concerning aardvarks which do not yet exist, or that there exist (now) fusions some parts (sub-states) of which do not (yet) exist. Perhaps none of these alternatives is actually self-contradictory or absurd, but the need to embrace one of them is scarcely an attractive feature of the standard account.

Were there no credible alternative accounts of what makes universal generalizations such as these true, we would have no option but to swallow these rebarbative consequences. But there *are* credible alternatives. Thus, in the case of our first example, it might be held that what makes it true that every natural number is immediately followed by another natural number, and what also makes true each individual instance of that general truth, is a conceptual connection – a connection between the concepts of *natural* or *finite number* and of *immediate succession*. Or – to canvas the alternative which I myself favour – we may hold that the general statement and its instances are made true by the essence or nature of the natural numbers, i.e. by what it is to be a finite or natural number. One reason to prefer an account along these lines is that it can better handle examples like my second than a conceptual connection account. That aardvarks are insectivores is a fact of nature, not something guaranteed by some connection between concepts. The point generalizes to many other universal statements to which the standard account appears ill-suited.

2.2 Accidental and lawlike generalizations

The contrast between general statements like *all my children live in England*, to which the standard account seems well-suited, and others like all aardvarks are insectivores, which seem to call for a different account is, at least roughly speaking, the contrast – already found in much work that pre-dates the emergence of truth-maker semantics – between what are often called 'accidental' and 'lawlike' generalizations. The former typically, and perhaps invariably, concern the elements of some fixed or closed finite totality which could, at least in principle, be specified piecemeal, whereas the latter are about an open class of things of some general sort which, even if finite, is not exhaustively specifiable.[9] Perhaps the most commonly

---

[9] It is perhaps not entirely clear that there could not be accidental generalizations about infinite totalities. It might be suggested, for example, that God could arbitrarily select an infinite subset of the natural numbers in such a way that, as it happens, the arabic numeral for each natural number in the subset include the numeral '7'. Nor is it quite obvious – even

emphasized difference between the two is that lawlike generalizations imply, or in some sense support, counterfactual statements, whereas merely accidental generalizations do not do so. Thus, to vary our example, given acceptance of a lawlike generalization such as *Common salt is soluble in water*, we may reasonably infer the singular counterfactual *Had this teaspoon of salt been immersed in water, it would have dissolved*. By contrast, given the merely accidental generalization *All the students in my logic class are male*, it would be rash, to say the least, to infer that *Had Alicia been in my logic class, she would have been male*.

In some relatively recent work, it has been argued that generalizations expressed in natural languages fall into at least two different groups, corresponding roughly with this familiar contrast, and that the differences are such as to warrant formalizing or regimenting them in different ways.[10] I shall refer to this as the *alternative logical form* view. It will, I think, help to forestall some possible misunderstandings if I say a little about how I think this view is related to my project.

We should note first that the class of general statements to which the standard account seems ill-suited is not happily labelled 'lawlike', given the very strong suggestion that label carries, that the generalization states a natural scientific law. So much is clear from the examples already discussed, since the fact that every natural number has another as its immediate successor is hardly to be regarded as a law of nature. Nor are the facts that all spinsters are unmarried, that anything red is coloured, that only people of age 18 and upwards are entitled to vote in Parliamentary Elections in the UK, and that everything is self-identical – but all, along with many others of the same sorts, are general statements for which the standard account is implausible. The more colourless 'non-accidental' would be a better label. This is not, however, a point of contrast with the view that different logical forms should be assigned to accidental and lawlike generalizations. Proponents of that view are likely to see the important difference between the two classes of general statement as lying in the capacity, or lack of it, to support

---

if it is tempting to think – that the number of things covered by a lawlike generalization, even if finite, will have no finite upper bound. Aardvarks are by nature insectivorous, but we can be pretty confident that there will never have been more than 10 billion of them.

[10]See, for example, Drewery (2005), developing ideas from Drewery (1998). As Drewery emphasizes, the contrast between the two sorts of generalization is generally well-marked syntactically in English.

What Makes True Universal Statements True?

counterfactuals, and it is clear that many general statements besides natural laws have that capacity.[11]

In essence, what the alternative logical form view claims is that while merely accidental generalizations such as *All the students in my logic class are male* are adequately represented as having the logical form of a universally quantified material conditional $\forall x(Fx \to Gx)$, non-accidental generalizations have a different, and probably more complex, logical form which cannot adequately be represented in standard first-order logic. There are different views about what is required to capture the logical form of non-accidental generalizations. Given that their main distinguishing feature is their capacity to support counterfactuals, it is natural to think that they involve some kind of modal element. This suggests that their logical form might be better represented in a modal extension of standard first-order logic. The simplest proposal would be to represent them as necessitated universally quantified material conditions, having the form $\Box \forall x(Fx \to Gx)$. But this oversimplifies in at least two respects. First, it seems clear that, while some kind of necessity may be involved in all non-accidental generalizations, different kinds of necessity – natural, mathematical, logical, legal, etc. – are in play in different examples. Second, not all generalizations which might be thought to call for different treatment are exceptionless, but, as Aristotle puts it, hold normally and for the most part. This goes for what are usually called *generics*, typically expressed in the form *Fs are Gs*, without any quantifier prefix such as 'all', such as *Kittens are born blind, Horses have four legs*, etc. The general truth that *aardvarks are insectivores* is not upset by the occasional aardvark tucking into a Big Mac. Drewery proposes that such generics might be taken to have the logical form $\Box \forall x((Fx \land Nx) \to Gx)$ where Nx says that x is 'non-exceptional in this case' (cf. Drewery, 2005, p. 383, drawing on Drewery, 1998). The additional strength of exceptionless generalizations might then, she suggests, be captured by strengthening to $\Box \forall x(((Fx \land Nx) \to Gx) \land (Fx \leftrightarrow Nx))$.[12] As Drewery observes, an alternative treatment of non-accidental generalizations may be given, deploying a special logic of sortal or kind terms.[13] It is unnecessary for me to explore these, or any other, specific proposals about logical form here. I mention them only to contrast them with my own proposal, which is not about logical form at all, but about what makes uni-

---

[11] This point is emphasized by Drewery (2005, pp. 380–81).
[12] Cf. Drewery (2005, p. 387).
[13] Cf. Drewery (2005, p. 384), where she refers to work by Lowe (1989, chs. 8,9) and others.

versally quantified statements true. As far as my proposal is concerned, we might as well regard all universal generalizations as having the simple logical form $\forall x A(x)$, with many of them exemplifying the more specific form of universally quantified material conditionals $\forall x(Fx \to Gx)$. The alternative logical form view is simply orthogonal to the concern which drives my proposal. Even if we were to agree that certain general statements are better represented as necessitated universally quantified conditionals, or in one of the more complicated ways mentioned above, this would leave the issue with which I am concerned pretty well untouched. For on any plausible alternative representation of its logical form, a non-accidental generalization (*all*) *Fs are Gs* will at least *entail* $\forall x(Fx \to Gx)$, and my question concerns what makes *that* entailed general statement true. Conversely, although I shall continue to assume that non-accidental general statements may be adequately represented as they are in standard first-order logic, my contention that what makes them true, when they are, differs from what makes merely accidental ones true does not carry any particular implications concerning their logical form.[14]

2.3 The structure of sentences and the structure of truthmakers

The standard account may seem to draw support from a plausible view about the relation between the structure of sentences and the structure of truthmaking states. It is both natural and plausible to view the sentences of a language – especially those of a regimented language – as either simple or complex, or at least as divisible into simpler and more complex sentences, with the more complex built up by means of various constructions from the simpler. Thus we have, at the bottom level as it were, simple or atomic sentences, composed out of names or singular terms and predicates. These may then be combined by means of sentence connectives or sentential operators to make various kinds of compound, such as negations, conjunctions, disjunctions, conditionals, and so on. Then there are quantified sentences, which may be viewed as in some sense composed out of their instances.

When we turn to the truth- or falsity-makers (or verifiers or falsifiers) for sentences of the various kinds, it is quite natural to suppose that these too have a kind of structure which runs parallel to the structure of sentences – so that sentences of the simplest kind, devoid of sentential operators and quantifiers, etc., are verified or falsified by states of a correspondingly simple kind, whereas the more complex sentences are made true by more complex

---

[14] But see fn. 21, page 19 below.

What Makes True Universal Statements True?

states which are, in some way, built up out of simpler states. In short, the structure of the states which serve as truth- or falsity-makers matches the structure of the sentences they make true or false – bottom up in both cases, from simpler to more complex.[15]

However natural this assumption may seem, it is certainly neither obviously correct – nor is it inevitable. It gains some plausibility from the equally, if not more, plausible belief that the truth- and falsity-makers for sentences should have a kind of compositional structure mirroring that of the sentences themselves. But one has only to look a little more closely at the compositional structure of sentences to see that it does not enforce the view that the states which verify universal propositions are strictly composed (say by fusion) out of the states which verify their instances, and leaves room for an alternative account. For while negative, conjunctive, and disjunctive sentences, for example, are quite literally composed of the simpler sentences of which they are the negations, conjunctions, or disjunctions, the same is not true of quantified sentences – $\forall x Bx$ is *not* literally composed of its instances $Ba_1, Ba_2, Ba_3, \ldots$. It is the result of applying the universal quantifier $\forall x \ldots x \ldots$ to the (typically complex) predicate $B(\xi)$.

The belief that $\forall x Bx$ is verified by a complex state composed of the simpler states which verify its instances separately – say by the fusion of the states $s_1 \sqcup s_2 \sqcup \ldots$ which make true the conjunction of each of its instances $Ba_1 \wedge Ba_2 \wedge Ba_3 \wedge \ldots$ – is a belief about the internal structure of the state which verifies the universal proposition, and cannot properly be evaluated without a closer consideration of the internal structure of states in general, including those states which verify atomic propositions.

It is certainly very plausible that among the simplest sentences of the language there will be sentences like 'Mary is asleep', 'Bill is taller than Jack', etc., and that the states which verify such sentences (or more precisely, the propositions such sentences may be used to express in particular contexts) will be in some sense made up of parts corresponding to the sub-sentential parts of these sentences – i.e. to the proper names or other singular terms in-

---

[15]This need not involve the idea that all states are either atomic, or are composed out of simpler, and ultimately out of atomic, states by fusion. Fine, emphasizing the abstractness and generality of his approach, explicitly distances himself from the assumption that 'all states are constructed from atomic states which are somehow isomorphic with the atomic sentences of the language under consideration' (Fine, 2017, p. 561). Later in the same paper he also stresses that there is no assumption that verifiers will be minimal (where a state minimally verifies a statement if it exactly verifies the statement but no proper part of the state does so) (Fine, 2017, §6).

volved, and to the predicates whose argument-places they occupy – and so have objects and properties or relations as constituents. Thus it is plausible that the proposition that Mary is asleep, if true, is made so by the obtaining of a state involving Mary and the property of being asleep – a state which consists, at least in part, in the instantiation of that property by that object.

It is plausible, too, that the states which verify or falsify compounds such as negations, conjunctions, and disjunctions, are structured in ways that match the structure of those kinds of proposition, so that a state s verifies $\neg A$ if it falsifies A, verifies $A \vee B$ if it verifies A or verifies B, and verifies $A \wedge B$ if it is composed of two sub-states s_A and s_B which verify A and B respectively.

When we turn to universal propositions, however, taking the internal structure of the sentences by which they are expressed as a guide to the internal structure of their truth- and falsity-makers does less to encourage the standard account. Consider, for example, the generalization that (all) cows are ruminants. If we regiment this sentence, as usual, as a universally quantified conditional, $\forall x(x\ is\ a\ cow \rightarrow x\ is\ a\ ruminant)$, its immediate constituents, besides the logical operators, are just the predicates $\xi\ is\ a\ cow$ and $\xi\ is\ a\ ruminant$. There is no definite reference to the individual objects which may be involved in the states which verify or falsify its instances. Of course, these individuals will belong to the domain over which the bound variable is taken to range. But the statement itself says nothing about which individuals that domain comprises. It would be entirely consistent with accepting the sentence's structure as a guide to the internal structure of its truth- and falsity-makers to take them to consist in higher-level relations (of inclusion and non-inclusion) holding between the properties of *being a cow* and *being a ruminant*.

3 Alternative exact truth-makers

Turning now to the questions I have been postponing, it will be useful to start with a more detailed – but still quite concise – account of exact truth-making, and some discussion of the requirements for exactness, before introducing an alternative account.

3.1 Standard exact truth-makers

In any form of truth-maker semantics, whether exact, inexact, or loose, the 'pluriverse' of possible worlds which forms the basis of world semantics is

What Makes True Universal Statements True?

replaced by a space of states, as the basis for the models of the language for which we give the semantics.[16]

A *state space* is a pair $\langle S, \sqsubseteq \rangle$, where S is a non-empty set (of states) and $\sqsubseteq$ is a partial order on S (i.e. $\sqsubseteq$ is reflexive, anti-symmetric and transitive).

A model $\mathfrak{M}$ for a first-order language $\mathcal{L}$ is a quadruple $\langle S, A, \sqsubseteq, |:| \rangle$, where $\langle S, \sqsubseteq \rangle$ is a state space, A is a non-empty set of individuals, and $|:|$ is a valuation function taking each n-place predicate F and any n individuals $a_1, a_2, \ldots, a_n$ to a pair $\langle \mathcal{V}, \mathcal{F} \rangle$ of subsets of S. Intuitively, $\mathcal{V}$ is the set of states which verify F of $a_1, a_2, \ldots, a_n$ and $\mathcal{F}$ the corresponding set of falsifiers. The former is also denoted by $|F, a_1, \ldots, a_n|^+$ and the latter by $|F, a_1, \ldots, a_n|^-$.

If we assume, as Fine (2017) does, that our language contains individual constants $\mathbf{a_1}, \mathbf{a_2}, \ldots$, one for each of the elements $a_1, a_2, \ldots$ of A, then *exact* verification and falsification clauses for atomic and complex statements may be stated as follows:[17]

(atomic)⁺ $s \Vdash F\mathbf{a_1}\ldots\mathbf{a_n}$ if $s \in |F, a_1, \ldots, a_n|^+$

(atomic)⁻ $s \dashv\vdash F\mathbf{a_1}\ldots\mathbf{a_n}$ if $s \in |F, a_1, \ldots, a_n|^-$

(¬)⁺ $s \Vdash \neg B$ if $s \dashv\vdash B$

(¬)⁻ $s \dashv\vdash \neg B$ if $s \Vdash B$

(∧)⁺ $s \Vdash B \wedge C$ if for some states t, u, $t \Vdash B$, $u \Vdash C$ and $s = t \sqcup u$

(∧)⁻ $s \dashv\vdash B \wedge C$ if $s \dashv\vdash B$ or $s \dashv\vdash C$

(∨)⁺ $s \Vdash B \vee C$ if $s \Vdash B$ or $s \Vdash C$

(∨)⁻ $s \dashv\vdash B \vee C$ if for some states t, u, $t \dashv\vdash B$, $u \dashv\vdash C$ and $s = t \sqcup u$

(∀)⁺ $s \Vdash \forall x \varphi(x)$ if there are states $s_1, s_2, \ldots$ with $s_1 \Vdash \varphi(\mathbf{a_1})$, $s_2 \Vdash \varphi(\mathbf{a_2})$, ... and $s = s_1 \sqcup s_2 \sqcup \ldots$

(∀)⁻ $s \dashv\vdash \forall x \varphi(x)$ if for some $a \in A$, $s \dashv\vdash \varphi(\mathbf{a})$

---

[16] The explanations which follow are taken from Fine (2017, §§4–7). There are some additional conditions on state spaces, but these will not be important here. To ease comparison, I have mostly retained Fine's notation. But I have used $\mathcal{V}, \mathcal{F}$ below, where Fine has plain V, F, to avoid possible confusion between the predicate letter F and the subset $F \subseteq S$.

[17] $s \Vdash A$ abbreviates 's verifies A' and $s \dashv\vdash A$'s falsifies A'.

(∃)⁺ $s \Vdash \exists x \varphi(x)$ if for some $a \in A$, $s \Vdash \varphi(\mathbf{a})$

(∃)⁻ $s \dashv \exists x \varphi(x)$ if there are states $s_1, s_2, \ldots$ with $s_1 \dashv \varphi(\mathbf{a_1})$, $s_2 \dashv \varphi(\mathbf{a_2}), \ldots$ and $s = s_1 \sqcup s_2 \sqcup \ldots$

Our question is whether, without forsaking the framework of exact truth-making, the clauses (∀)⁺ and (∃)⁻ might be modified to allow for verification of $\forall x \varphi(x)$ and falsification of $\exists x \varphi(x)$ by generic states. We should, accordingly, first get clear what exactness requires.

3.2 Exactness

As we have seen, Fine takes a state s to be an exact verifier for a statement p if and only if it is *wholly* relevant to p, and to be an inexact verifier if it is *partially* relevant.[18] But how, precisely, is 'wholly relevant' to be understood in this context? Contrasting exact with inexact verification, Fine gives the example:

> The presence of rain will be an exact verifier for the statement 'it is rainy'; the presence of wind and rain will be an inexact verifier for the statement 'it is rainy', though not an exact verifier. (Fine, 2017, p. 558)

This might be taken to suggest that a state cannot be an exact verifier for (and so be wholly relevant to) a statement if it has a proper part which does not verify the statement (as, presumably, the state of its being rainy and windy has the state of its being windy as a proper part which is not a verifier for 'it is rainy'). But this cannot be what Fine intends. For, after emphasizing that 'state' is for him a term of art which need not stand for a state 'in any intuitive sense of the term', he writes:

> It should be noted that our approach to states is highly general and abstract. We have formed no particular conception of what they are; and nor have we assumed that there are 'atomic' states, from which all other states can be obtained by fusion. (Fine, 2017, p. 561)

---

[18]Cf. Fine (2017, p. 558). Fine (in press) says that an exact verifier must be 'relevant as a whole' to the statement it verifies. I think we should understand both these words and 'wholly relevant' as expressing the idea that the whole of the state s is relevant to verifying the statement p.

What Makes True Universal Statements True?

Not only is it not assumed that there are atomic states from which all other states are obtainable by fusion, but it is not assumed that exact verifiers will be *minimal* – where 'the state s minimally verifies the formula A if s exactly verifies A and if no proper part of s exactly verifies A (i.e. if $s' \sqsubseteq s$ and $s' \Vdash A$ implies $s' = s$)' (Fine, 2017, p. 564). Fine (loc. cit.) gives the following example:

> Now suppose that p is the sole verifier of p and q the sole verifier of q, with $q \neq p$. Then p and $p \sqcup q$ are both verifiers of p∨(p∧q), with $p \sqcup q$ non-minimal since it contains the verifier p as a proper part.

This example shows that an exact verifier can have a proper part which does not verify the statement the exact verifier verifies – for $p \sqcup q$ is an exact verifier for p∨(p∧q), but it contains as a proper part the state q, which is *not* a verifier for p∨(p∧q). A little later, Fine (loc. cit.) adds this clarification:

> The relevant sense in which an exact verifier is wholly relevant to the statement it makes true is not one which requires that no part of the verifier be redundant but is one in which each part of the verifier can be seen to play an active role in verifying the statement. Thus the verifier $p \sqcup q$ of p ∨ (p ∧ q) can be seen to play such an active role, even though the part q is redundant, because of its connection with the second disjunct (p ∧ q).

It would be curmudgeonly to complain that 'play an active role in verifying a statement' lacks precision – Fine's intention is clear enough. Where $s = t \sqcup u$ and t (exactly) verifies B but not C, s has a part, t, which plays no active role in verifying C, so that s is not wholly relevant to C, and does not exactly verify it, although it does so inexactly. This is in clear contrast with $p \sqcup q$ in Fine's example, where the part q does play an active role in verifying p∨(p∧q), being an indispensable part of a verifier for the right disjunct p∧q.

If being wholly relevant, and hence exactness, is understood in this way, then it is clear that there is no reason why one and the same state should not be an exact verifier of more than one statement. In particular, it is not ruled out that there should be a single state which has no proper parts (i.e. is not the fusion of simpler states) and which exactly verifies a universally quantified statement $\forall x \varphi(x)$ and also exactly verifies each of its instances $\varphi(a_1), \varphi(a_2), \ldots$, i.e a state of precisely the kind required for *generic* verification or falsification. I conclude that there is no bar to generic verification or falsification inherent in the conception of exact truth- and falsity-making.[19]

3.3 Modified exact truth-makers

Whilst the idea that universal statements may be rendered true by general connections between the properties involved, rather than made true piecemeal by states which verify their instances, seems straightforward enough, it is a further – and much less straightforward – question how to implement this idea in the framework of exact truth-maker semantics. This is largely because we must to some extent depart from the bottom-up determination of truth-values which is built into truth-maker semantics as we have it, and which – notwithstanding its divergence in other respects – it takes over from standard model theoretic semantics in general. In standard first-order semantics, the determination of truth-values for all sentences is driven by an initial assignment of individual objects from the given domain, D, to whatever simple, non-logical predicates the language may contain. That is, we have a function, v, which assigns to each simple n-place predicate F a subset of D^n. This determines, together with an assignment of elements of D to any constant terms and perhaps with the help of an assignment to individual variables, the truth-values of closed and open atomic sentences, and thence, via the clauses for the connectives and quantifiers, those of the complex sentences of the language. In some variations on this standard semantics, predicates may be assigned both extensions and counter-extensions, exclusive of one another, but not necessarily jointly exhaustive

---

[19] I don't expect Fine to disagree with this conclusion, since at the end of (Fine, 2017, pp. 568–569), he writes: 'it might be thought that $\forall x \varphi(x)$ is verified, in the first place, by certain general facts which, in themselves, do not involve any particular individuals. It turns out that this idea of generic verification can be developed within the framework of arbitrary objects developed in Fine (1985). Thus the verifier of "all men are mortal" might be taken to be the generic fact that the arbitrary man is mortal.' There is no suggestion that adopting this course would be to abandon exact verification. The proposal I shall develop certainly differs from the alternative Fine envisages here, since it makes no use of arbitrary objects or states involving them, but I can see no reason why it should be thought to sacrifice exactness.

In another paper Fine (in press) suggests that for *inexact* semantics, the clause for verification of universal statements might run: a state is a truth-maker for the universal quantification $\forall x B(x)$ iff it is a truth-maker for each of $B(\mathbf{a_1}), B(\mathbf{a_2}), \ldots$ The implication, in context, is that this clause does not specify an exact truth-maker. This does not go against the claim made in the text, as it might at first appear to do. I am not claiming that just *any* state which verifies each of the instances of a universal quantification will be an exact verifier – even if s exactly verifies each of $B(\mathbf{a_1}), B(\mathbf{a_2}), \ldots$, its fusion with a state t which is irrelevant to them and to $\forall x B(x)$ will be an inexact verifier for the latter by Fine's clause, but not an exact verifier.

What Makes True Universal Statements True?

of the individual domain — but the overall pattern of truth-value determination remains firmly bottom-up, driven by initial assignments of individuals to predicates. First-order truth-maker semantics differs in working with a domain of states as well as a domain of individuals, and its assignment function |:| assigns states to simple n-place predicates relative to n-tuples of individuals — $|F, a_1, \ldots, a_n|^+$ comprising the states, if any, which make F true of $a_1, \ldots, a_n$, and $|F, a_1, \ldots, a_n|^-$ comprising the states, if any, which make F false of $a_1, \ldots, a_n$. But bottom-up determination remains in place.

The salient point, for our present purposes, is that in each type of semantics, the initial assignments — whether they be of individuals to the extensions of simple predicates, or individuals to the extensions and counter-extensions of such predicates, or of verifying and falsifying states to pairs of predicates and n-tuples of individuals — are entirely independent of one another. This may encourage, even if it does not strictly imply, a certain sort of metaphysical atomism — any relations *between* predicates (or better, between the properties and relations for which they stand) are pictured as *consequential* upon underlying independent facts about which individuals happen to lie in the extensions of those predicates taken separately. Of course, it is essential that the range of admissible models should cover all the logical possibilities. But as we shall see — if it is not already clear enough — that need not preclude extending the assignment function in ways which capture the intuitive idea that relations between properties, and thereby, the truth-values of some statements, may be determined, not at the bottom-level, but at a higher, generic level.

One way — perhaps the most obvious way — to accomplish this would be to extend the function |:| so as to allow, as well as assignments of states to predicate/n-tuple pairs, further assignments to predicate/predicate pairs. Thus where F and G are predicates of the same arity, we might have the assignments $|F, G|^+$ and $|F, G|^-$ — where the former comprises those states which verify G of any individuals $a_1, \ldots, a_n$ of which F is true (i.e. for which $|F, a_1, \ldots, a_n|^+ \neq \varnothing$), and the latter those which falsify G of any such individuals. Intuitively, we might think of a state in $|F, G|^+$ as a kind of generic state which simultaneously verifies both the universal statement $\forall x(F(x) \rightarrow G(x))$ and each of its instances $F(\mathbf{a_1}) \rightarrow G(\mathbf{a_1}), F(\mathbf{a_2}) \rightarrow G(\mathbf{a_2}), \ldots$

How should we formulate an appropriately modified clause for quantified statements? I think that before we can answer this question, we need to see how to generalize the idea just presented to accommodate a wider range of forms of generalization. To see how we may do so, notice first that an extension of |:| which has the same effect as extending it to a predicate pair $\langle F, G \rangle$ in the way proposed would be to allow its application to the open statement $F(x) \to G(x)$, with $|F(x) \to G(x)|^+$ and $|F(x) \to G(x)|^-$ defined to comprise, respectively, those states which verify and those which falsify the open statement. The obvious advantage of the shift – effectively largely notational – is that this is readily generalized. We simply allow application of |:| to any open statement. We may then modify the verification clause for universal quantification as follows:

∀⁺ $s \Vdash \forall x B(x)$ iff either $s \in |B(x)|^+$ or $s = s_1 \sqcup s \sqcup_2 \ldots$ where $s_1 \Vdash B(\mathbf{a_1}), s_2 \Vdash B(\mathbf{a_2}), \ldots$

Note that the falsification clause remains unchanged:

∀⁻ $s \dashv\Vdash \forall x B(x)$ iff $s \dashv\Vdash B(\mathbf{a})$ for some a

Should there be a state $s \in |B(x)|^-$, s verifies $\forall x \neg B(x)$ – i.e. the contrary of $\forall x B(x)$, as opposed to its contradictory.

There are, however, some changes to the clauses for the connectives consequential upon the possibility of generic verification. For example, if a generic state verifies $B(x) \vee C(x)$, where x is the sole free variable, it will also verify the specific closed formulae $B(\mathbf{a_7}) \vee C(\mathbf{a_7})$, where $\mathbf{a_7}$ replaces free x throughout. Accordingly, we should amend (∨)⁺ to:

(∨)⁺* $s \Vdash B \vee C$ if $s \Vdash B(x) \vee C(x)$ or $s \Vdash B$ or $s \Vdash C$

A similar adjustment is required for the verification clause for the conditional.[20]

An extension of (exact) truth-maker semantics along these lines is, in a certain sense, the *minimum* required to make space for generic verification. It allows for models in which some general statements and their instances are generically verified, whilst leaving room for instance-by-instance verification for others. Of course, a simplification of the semantics would be possible, were we to exclude the possibility of instantial verification of universal statements. On the other hand, an account which leaves room for both

---

[20]No change is needed in the verification clause for conjunction. For a generic state g which verifies $B(x) \wedge C(x)$ no matter which element of the individual domain x takes as its value will verify each of $B(x)$ and $C(x)$ for any element as taken x's value, and – since generic states have no stately parts – will verify each of them exactly. The verifier $s = t \sqcup u$ for $B \wedge C$ may therefore be the fusion $g \sqcup g$ – there is no requirement in the clause for conjunction that t and u be distinct.

kinds of verification is more realistic, in the sense that it recognizes that merely accidental generalizations are verified piecemeal, in contrast with non-accidental ones verified generically.[21]

References

Barwise, J., & Cooper, R. (1981). Generalized quantifiers and natural language. *Linguistics and Philosophy, 4,* 159–219.
Barwise, J., & Perry, J. (1983). *Situations and Attitudes.* Cambrige: MIT Press.
Davidson, D. (1967). Truth and meaning. *Synthese, 17,* 304–323.
Drewery, A. (1998). *Generics, Laws, and Context* (Unpublished doctoral dissertation). University of Edinburgh.
Drewery, A. (2005). The logical form of universal generalizations. *Australian Journal of Philosophy, 83,* 373–793.
Fine, K. (1985). *Reasoning with Arbitrary Objects.* Oxford: Blackwell.
Fine, K. (2017). Truthmaker semantics. In B. Hale, C. Wright, & A. Miller (Eds.), *A Companion to the Philosophy of Language* (pp. 556–577). Blackwell.
Fine, K. (in press). The world of truthmaking. In I. Fred & J. Leech (Eds.), *Being Necessary: Themes of Ontology, Modality, and the Reations between them.* Oxford University Press.
Hale, B. (2013). *Necessary Beings: An Essay on Ontology, Modality, and the Relations between them.* Oxford: Oxford University Press.

---

[21] It is arguable that such an account is best developed in a framework which replaces the standard quantifiers ∀ and ∃ by generalized quantifiers of the kind proposed by Barwise and Cooper (1981). In particular, taking quantifiers to be semantically complex expressions formed by applying a determiner ('every', 'some', 'no', ...) to a common noun(-phrase) – so that a generalization such as 'Every aardvark is insectivorous' is represented as the result of applying a restricted quantifier 'every aardvark' to a predicate '... is insectivorous' – allows us to see the distinction between accidental and non-accidental generalizations as grounded in different kinds of restriction implied by the common noun to which the determiner is attached to form the initial quantifier. Very roughly, while nearly all common noun(-phrase)s serve to restrict the range of quantifiers in which they are embedded, some – examples are 'person in this room' and 'member of my logic class' – imply a bound on the size of that range; it is in such cases that we have an accidental generalization. The distinction between bounding and non-bounding restrictors is not usually marked syntactically, however, and often relies upon contextual factors. In consequence, it is not straightforwardly implementable for a formal language and its semantics. I must postpone further development of these ideas to another occasion. I am grateful to Ian Rumfitt for urging on me the advantages of the generalized quantifier approach.

Humberstone, I. L. (1981). From worlds to possibilities. *Journal of Philosophical Logic*, *10*, 313–339.
Lowe, E. J. (1989). *Kinds of Being*. Oxford: Blackwell.
Rumfitt, I. (2015). *The Boundary Stones of Thought*. Oxford: Clarendon Press.

Choosing Your Nonmonotonic Logic: A Shopper's Guide

ULF HLOBIL[1]

Abstract: The paper presents an exhaustive menu of nonmonotonic logics. The options are individuated in terms of the principles they reject. I locate, e.g., cumulative logics and relevance logics on this menu. I highlight some frequently neglected options, and I argue that these neglected options are particularly attractive for inferentialists.

Keywords: Nonmonotonic logic, Inferentialism, Cumulative transitivity, Cautious monotonicity, Relevance logic

Nonmonotonic logics are logics in which Weakening or Monotonicity (MO) sometimes fails.

MO $\qquad$ If $\Gamma \vdash A$, then $\Gamma, \Delta \vdash A$.

Given that you are reading this, you are presumably in the market for a nonmonotonic logic. In this paper, I will offer some guidance by giving you a complete menu of options. The different options are individuated by the principles they reject. I will highlight some frequently neglected options that are attractive for inferentialists.

I use the snake-turnstile "$\mid\!\sim$" to talk about nonmonotonic consequence. I take permutation and contraction for granted by working with sets on the left (and the right, in multiple conclusion logics) of the turnstile.

1 What do you need?

In order to know which nonmonotonic logic is right for you, we must know what you want your nonmonotonic logic *for*. There are two families of reasons for wanting a nonmonotonic logic. Either you want to get more

---

[1] I would like to thank Robert Brandom, Daniel Kaplan, Shuhei Shimamura, Rea Golan, and everyone who supported the Pittsburgh research group on nonmonotonic logic over the years.

conclusions than classical logic gives you, or you want to get fewer (or both). The logics that are typically called "nonmonotonic logics" give us more conclusions than classical logic.[2] They add risky inferences, like the following:

(a) Zazzles is a cat. So Zazzles has four legs.

(b) I let go of this object h meters above ground. So it will hit the ground with $\sqrt{2gh}\ \frac{mtr}{sec}$.

These inferences are obviously not classically valid; and they are defeasible. If we add to the first one the premise that Zazzles has (tragically) lost a leg in an accident, the conclusion no longer follows. And if we add to the second inference the premise that the object is a bird, the conclusion no longer follows. Nevertheless, the conclusions intuitively follow from the premises. Nonmonotonic logics in the first family try to capture this intuition.

Let us now turn to the second kind of motivation. Relevance logicians[3] reject MO in order to avoid fallacies of relevance, like the result that "It is not the case that if you are a philosopher, you are dumb" entails "If you are dumb, the moon is made of cheese." To avoid such results, relevance logicians restrict the classical principle that a set entails its elements (containment or CO), while accepting that every sentence entails itself (reflexivity or RE).

CO   If $A \in \Gamma$, then $\Gamma \mid\!\sim A$.

RE $A \mid\!\sim A$.

In the atomic case, e.g., the set $\{p, q\}$ doesn't entail p. For relevance logicians, this is justified because q is not relevant to p, where the notion of relevance is usually spelled out in terms of variable sharing.

Given these two families of motivations, any shopping tour should start with the question: Do you care about allowing risky inferences? Or do you want to avoid fallacies of relevance?

If you merely want to avoid fallacies of relevance and don't care about risky inference, I don't have any new insights to offer. You should shop around to find a relevance logic you like (see Anderson & Belnap, 1975; Dunn & Restall, 2002). If you care about both, relevance and risky inferences, I may have something new for you (see NM-LR below).

---

[2] Sometimes some other monotonic logic is used as the lower limit.
[3] For my present purposes, linear logic and similar logics count as relevance logics.

Shopper's Guide: Nonmonotonic Logic

If you want to codify risky inferences and don't care about relevance, there are many logics on the market that you might like, e.g., preferential logics (Kraus, Lehmann, & Magidor, 1990), default logics (Reiter, 1980), adaptive logics (Batens, 2007), argument-based approaches (Dung, 1995), etc. In fact, there is a confusing variety of such logics. They are often categorized according to the technical machinery they use (Strasser & Antonelli, 2016). Preferential logics, e.g., use partial orders over worlds. Default logics use default rules. Adaptive logics use sets of abnormalities. Argument-based approaches use graphs in which the nodes are arguments. This is often not very helpful because you may not care what technical machinery is running under the hood of your logic.

I want to look at nonmonotonic logics from a more abstract perspective. This will allow us to explore the logical space in which nonmonotonic logics are located in terms of philosophical views about logic and principles you may want your logic to satisfy, where this includes structural principles and principles governing connectives.

2 The full menu

You cannot get a nonmonotonic logic without having to give up some principles that many find desirable. The good news is that you get a choice regarding which principles you want to give up. In this section, I will go through some of these choices. The result will be an exhaustive (but not exclusive) classification of nonmonotonic logics into seventeen types.

We can think of these seventeen types as generated by four choices. At each choice point, you must reject at least one of a given set of principles. The first choice (labeled [C1] in Figure 1) is that between rejecting CO and rejecting Mixed-Cut.

Mixed-Cut If $\Gamma \hspace{-0.5ex}\mid\hspace{-1.2ex}\sim A$ and $\Delta, A \hspace{-0.5ex}\mid\hspace{-1.2ex}\sim B$, then $\Gamma, \Delta \hspace{-0.5ex}\mid\hspace{-1.2ex}\sim B$.

You must choose between these two because CO together with Mixed-Cut implies MO. For, since by CO $B, A \hspace{-0.5ex}\mid\hspace{-1.2ex}\sim B$, if we have $\Gamma \hspace{-0.5ex}\mid\hspace{-1.2ex}\sim A$, we get $\Gamma, B \hspace{-0.5ex}\mid\hspace{-1.2ex}\sim A$ by Mixed-Cut. If we reject CO, we are in the area of relevance logics (in a very broad sense). If we reject Mixed-Cut, we reach our next choice point.

The second choice (labeled [C2] in Figure 1) is that between rejecting CO or the Deduction-Detachment Theorem (DDT) or Cumulative Transitivity (CT, aka Cut).

DDT $\Gamma \hspace{-0.5ex}\mid\hspace{-1.2ex}\sim A \to B$ iff $\Gamma, A \hspace{-0.5ex}\mid\hspace{-1.2ex}\sim B$.

CT   If $\Gamma\mathrel{\mid\!\sim} A$ and $\Gamma, A\mathrel{\mid\!\sim} B$, then $\Gamma\mathrel{\mid\!\sim} B$.

We cannot have all three because, by CO, $\Gamma, A, B\mathrel{\mid\!\sim} A$ and so, by DDT, $\Gamma, A\mathrel{\mid\!\sim} B \to A$. This means that we can get $\Gamma, B\mathrel{\mid\!\sim} A$ from $\Gamma\mathrel{\mid\!\sim} A$ by CT and (left to right) DDT (see Arieli & Avron, 2000).

The third choice (labeled [C3] in Figure 1) is one between rejecting the principle that everything implies all instances of the law of excluded middle (PEM), Cautious Monotonicity (CM), and a principle that I shall call Premise Fission (PF), which is basically reasoning by cases (as an invertible metarule).

PEM $\Gamma\mathrel{\mid\!\sim} A \vee \neg A$.

CM   If $\Gamma\mathrel{\mid\!\sim} A$ and $\Gamma\mathrel{\mid\!\sim} B$, then $\Gamma, B\mathrel{\mid\!\sim} A$.

PF $\Gamma, A \vee B\mathrel{\mid\!\sim} C$ iff $\Gamma, A\mathrel{\mid\!\sim} C$ and $\Gamma, B\mathrel{\mid\!\sim} C$.

To see that we must reject one of these, suppose again that $\Gamma\mathrel{\mid\!\sim} A$. By PEM, $\Gamma\mathrel{\mid\!\sim} B \vee \neg B$. By CM, $\Gamma, B \vee \neg B\mathrel{\mid\!\sim} A$. By PF, $\Gamma, B\mathrel{\mid\!\sim} A$.[4]

The fourth choice (labeled [C4] in Figure 1) concerns conjunction and disjunction. We must reject either PF, or Distribution (DI), or what I call Premise Fusion (FU).

DI   If $\Gamma, A \vee (B\&C)\mathrel{\mid\!\sim} D$, then $\Gamma, (A \vee B)\&(A \vee C)\mathrel{\mid\!\sim} D$.

FU $\Gamma, A\&B\mathrel{\mid\!\sim} C$ iff $\Gamma, A, B\mathrel{\mid\!\sim} C$.

PF requires an additive disjunction left-rule. FU requires a multiplicative conjunction left-rule. An additive disjunction doesn't distribute over a multiplicative conjunction (on the left of the turnstile).[5] To see this, notice that if $\Gamma, A\mathrel{\mid\!\sim} D$ and $\Gamma, B, C\mathrel{\mid\!\sim} D$, we get $\Gamma, A \vee (B\&C)\mathrel{\mid\!\sim} D$ by FU and PF. But if also $\Gamma, B, A\mathrel{\mid\!\not\sim} D$, we cannot get $\Gamma, (A \vee B)\&(A \vee C)\mathrel{\mid\!\sim} D$.

We can use these four choice points to give an exhaustive categorization of nonmonotonic logics. If we suppress some of our choices for relevance logics (by leaving out the CM, PEM, PF choice), we can distinguish seventeen types of nonmonotonic logics. They correspond to the leaves of the tree in Figure 1.

---

[4]This argument was brought to my attention by Dan Kaplan. It is used in the literature, e.g., by Arieli and Avron (2000).

[5]More precisely, an additive disjunction distributes over a multiplicative conjunction across the turnstile. In a multiple conclusion setting, we also get: If $\Gamma\mathrel{\mid\!\sim} \Delta, A \vee (B\&C)$, then $\Gamma\mathrel{\mid\!\sim} \Delta, (A \vee B)\&(A \vee C)$. However, we don't get DI.

Shopper's Guide: Nonmonotonic Logic

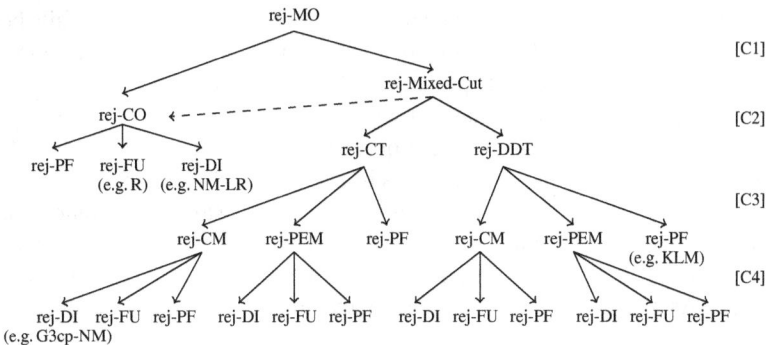

Figure 1: Types of nonmonotonic logic

Figure 1 should be read as follows: Every nonmonotonic logic must reject all the principles that occur on at least one complete branch of the tree. We can label the seventeen types by the principles that they reject. The relevance logic R, e.g., belongs to the type *Rej(CO,FU)*. For CO doesn't hold in R and the conjunction of R doesn't obey FU (even though the fusion operator does).[6] The dashed arrow indicates that the choice between rejecting CT and DDT is only forced if you accept CO. In fact, logics like NM-LR below obey CT and DDT; this is possible because CO doesn't hold in NM-LR.

Of course, a logic can always reject more principles than what the tree in Figure 1 requires. Hence, a logic can belong to several of our seventeen types. The logic NM-LR below, e.g., belongs to *Rej(CO,DI)* and *Rej(CO,FU)* because it rejects DI and FU. These types are, however, exhaustive in the sense that every nonmonotonic logic must belong to at least one of our seventeen types.

Let me illustrate the use of this categorization. Cumulative logics, in the sense of KLM (Kraus et al., 1990), accept CO and CT. This puts cumulative logics in the branch that rejects DDT. Cumulative logics also accept CM, and they are supra-classical, which means that they must accept PEM. So, cumulative logics belong to the type: *Rej(Mixed-Cut,DDT,PF)*. That tells us that Premise Fission cannot hold in cumulative logics.

---

[6]Once we have different kinds of conjunction (and perhaps disjunction) around, there is a question what principles like PF, FU, and DI mean. The answer is that what matters is just that PEM, PF and DI use the same disjunction and FU and DI use the same conjunction.

In preferential logics, which are a kind of cumulative logic, PF fails because we get $\Gamma, A \vee B \mid\!\sim C$ in cases in which $\Gamma, A \mid\!\not\sim C$. As an example, suppose that, in our favorite preferential logic, "Zazzles is a cat" nonmonotonically implies "Zazzles has four legs." By supra-classicality "Zazzles is a cat" implies "Zazzles either did or didn't have an accident in which he lost a leg." By CM, the set {"Zazzles is a cat", "Zazzles either did or didn't have an accident in which he lost a leg"} implies "Zazzles has four legs." That seems counter-intuitive. After all, "Zazzles is a cat. And Zazzles had an accident in which he lost a leg" clearly doesn't imply "Zazzles has four legs." As is clear from Figure 1, cumulative logics cannot avoid such cases.

3 Motivating choices

The tree in Figure 1 gives us a complete menu of options. Some of these options have well-known representatives, like KLM or the logic R. Below I will focus on options that have not been explored in the literature. I will present one logic of type *Rej(Mixed-Cut,CT,CM,DI)*, which I call G3cp-NM, and one logic of type *Rej(CO,DI)*, which I call NM-LR. The latter logic stands to LR as G3cp-NM stands to classical logic. Before we get into any technical details, however, we should reflect on our menu of options from a philosophical perspective.

The choices that KLM makes, i.e. *Rej(Mixed-Cut,DDT,PF)*, can be seen as embodying two priorities:[7] (a) rejecting as few structural principles as possible, and (b) staying supra-classical. To reject CM or CT would be to reject a structural principle where you could instead reject principles about particular connectives, namely DDT and PF respectively. So this would go against priority (a). Rejecting PEM instead of PF would preclude KLM from being supra-classical and, hence, go against priority (b).

Logics of the type *Rej(Mixed-Cut,CT,CM,DI)*, like the logic NM-G3cp below, can be seen as embodying the idea that we shouldn't reject principles regarding the behavior of connectives if we can instead reject structural principles. One way to flesh out this view is to say that satisfying DDT is part of what it means for something to be a conditional. Similarly, satisfying PF is part of what it means for something to be a disjunction. If we also want to stay supra-classical and, hence, accept CO and PEM, then it is al-

---

[7]Of course, I don't want to do the KLM advocate's introspection for her. And there can be other priorities that might motivate the choices KLM makes. But (a) and (b) are at least a natural interpretation.

ready settled that we must reject CT and CM. If we now add that satisfying FU is part of what it means to be a conjunction, we arrive at a logic of type *Rej(Mixed-Cut,CT,CM,DI)*.

This comparison brings out that navigating our decision-tree can be seen as an exercise in trading off structural principles that one may take to be constitutive of consequence against principles that one may consider constitutive of something being a certain connective. If you think, e.g., that CO and CT, but not CM, are constitutive of consequence and that FU isn't constitutive of conjunction but PF and PEM are constitutive of disjunction, then *Rej(Mixed-Cut,DDT,CM,FU)* is probably a nonmonotonic logic that you should consider.

If you are a logical inferentialist, you think that something has the meaning of, say, a conditional just in case, and because, it has a particular inferential role. In that case, you may plausibly think that principles like DDT, PF, and FU are constitutive of our connectives meaning what we want them to mean. For these are principles that plausibly characterize a part of the inferential roles of the conditional, disjunction and conjunction, respectively. Such an inferentialist view seems attractive to me and, hence, I will focus on options that are attractive from an inferentialist perspective. Before we get to this, however, I want to point out another issue for which inferentialist inclinations are relevant for the choice of a nonmonotonic logic, namely what we think logic should do for us.

4 Sticking to logic

If you are an inferentialist, it is probably not only the extension of your nonmonotonic consequence relation that will matter to you but also the way it is generated. You will probably say that the inferential roles of logically complex sentences are determined by the inferential rules governing the logical connectives, together with the inferential roles of the atomic sentences. Hence, you will want your nonmonotonic consequence relation to be generated by putting together a consequence relation over atomic sentences and rules governing the inferential behavior of the logical connectives. In this section, I will discuss implications of this view for the choice of a nonmonotonic logic.

We want to codify inference like that from "Zazzles is a cat" to "Zazzles has four legs." Hence, our consequence relation cannot be formal, in the sense of being closed under substitution (see Makinson, 2003). This isn't surprising. The inference about Zazzles has something to do with cats and legs and the meanings of "cat" and "legs." Substantive information about cats and legs and the meanings of "cat" and "legs" is curled up in our inference about Zazzles.

The logician has no particular expertise regarding cats and legs or the meanings of "cat" and "legs." So in order to construct a consequence relation in which "Zazzles has four legs" (or a sentence that stands for it in an artificial language) is a consequence of "Zazzles is a cat" (or a sentence that stands for it in an artificial language), the logician must draw on extra-logical knowledge. In particular, the atomic fragment of our consequence relation must be given to the logician from elsewhere. The logician gets no say in determining it. After all, implication relations among atomic sentences always embody substantive connections regarding which the logician cannot claim any expertise.

Of course, the logician may only be interested in atomic consequence relations that have certain structural properties, like CO, RE, CT or CM. But in order to justify such a restrictions from a philosophical perspective, we must give a philosophical interpretation of consequence, i.e. of the turnstile, and we must motivate the idea that all atomic consequence relations have certain properties on the basis of that interpretation.

This point is important because many nonmonotonic logicians are motivated by problems in artificial intelligence. For the purposes of artificial intelligence, however, it doesn't matter when and how we bring in extra-logical information, as long as we solve the practical problems at hand. We may, e.g., want a machine to deduce as much useful and reliable information as possible, in whatever way possible.

So in choosing a nonmonotonic logic, you should ask yourself: Do you allow your logic to partly determine the atomic fragment of your consequence relation? Or do you want a logic that conservatively extends any consequence relation over atomic sentences? If you want a maximally powerful inference engine, the first choice is preferable. If, however, you want your logic to determine the inferential roles of logically complex sentences in terms of the inferential roles of simpler sentences, then the second choice is preferable.

Comparing default logic and KLM can help to clarify my point. In default logic, we encode extra-logical information into default rules, which allow us to add a conclusion to our current knowledge base if we can get the so-called prerequisite and all the so-called justifications are consistent with our current knowledge base. We then apply our default rules until we reach a fixed point.⁸ Suppose, e.g., we know that Zazzles is a cat and that Zazzles likes to sleep on the sofa. Our only default rule is one that allows us to infer that something is hairy if it is a cat and it is consistent with our knowledge base that Zazzles is hairy. According to default logic, we can infer that Zazzles is hairy. And we can do that without ever explicitly feeding our logic the information that {"Zazzles is a cat", "Zazzles likes to sleep on the sofa"} nonmonotonically implies "Zazzles is hairy." Thus, default logic partly determines the consequence relation over atomic sentences.

Preferential semantics, in the spirit of KLM, differs in this respect from default logic. In KLM, we start with a partial order over states. We say that $\Gamma \mathrel{\mid\!\sim} A$ iff all the minimal states in which all members of Γ are true are also states in which A is true. We can think of the partial order as representing the extra-logical information which we are feeding into our cumulative logic. This ordering can be given antecedently to the semantic clauses that embody an account of the logical constants. If we interpret KLM in this way, we say, in effect, that all the extra-logical information must be given before we determine the consequence relation over logically complex sentences.⁹ Where KLM diverges from inferentialism is in giving a model-theoretic account of the meanings of the logical constants.

The question whether you want a nonmonotonic logic that is suitable for purposes in artificial intelligence or one that is compatible with inferentialism isn't a question about the extension of your consequence relation. Rather, it is a question about what you take to be basic. If you are an inferentialist, you should take nonmonotonic consequence relations over atomic sentences to be basic. Given that nonmonotonic logics that are suitable for artificial intelligence are well known, I want to close by presenting two options that are attractive from an inferentialist perspective—both in the extension of their consequence relations and in how they are generated.

5 Two inferentialist nonmonotonic logics

In this section, I will present two nonmonotonic logics that serve as examples of what I consider inferentialism-friendly nonmonotonic logics. I have

---

⁸That procedure defines an extension of our knowledge base. Skeptical consequences of our knowledge base are then those sentences that are in every extension. And credulous consequences are those that are in at least one extension.

⁹That is why, in KLM, "Nixon is a pacifist" does not follow from "Nixon is a Quaker and Nixon is a Republican", unless we order the states in such a way that in all the minimal states in which Nixon is a Quaker and a Republican, he is also a pacifist (Kraus et al., 1990).

presented a similar logic elsewhere (Hlobil, 2016); so I will focus on what is new and skip details that can be found in previous presentations of the framework.

5.1 The basics

Let $\mathfrak{L}_0$ be a set of atomic sentences, and let $\mathfrak{L}$ be the result of adding the connectives $\rightarrow, \vee,$ and $\&$ (and $\neg$ in the second logic; it can be treated as defined in the first logic) to this atomic language, with their usual syntax.

Let's start with a (multiple conclusion) nonmonotonic consequence relation over atoms.[10]

Definition 1 *A base consequence relation,* $\mid\!\sim_0\,\subseteq \mathcal{P}(\mathfrak{L}_0) \times \mathcal{P}(\mathfrak{L}_0)$, *is a relation between sets of atomic sentences.*

This definition is maximally liberal. In the context of a particular logic, we will add constraints. If you are an inferentialist, you will think that such base consequence relations define (at least in part) the meanings of the atomic sentences. We will use our base consequence relation as a set of axioms in otherwise familiar sequent calculi. The resulting consequence relation over the whole language is $\mid\!\sim$.

5.2 Making G3cp nonmonotonic

Suppose we want a supra-classical nonmonotonic logic. In that case, we should choose sequent rules for classical logic. However, we don't want to enforce structural principles like monotonicity. Hence, we should choose a sequent calculus for classical logic in which Weakening is absorbed.

The sequent calculus G3cp fits the bill (Troelstra & Schwichtenberg, 2000). The axioms of G3cp are $\Gamma, p \vdash p, \Delta$ and $\bot, \Gamma \vdash \Delta$. Like in LK, it suffices in G3cp to use just the atomic instances of these axioms to get classical logic. So if we make sure that all of our base relations include all atomic instances of the axioms of G3cp, our resulting nonmonotonic logic will be supra-classical. Let's say that base relations that include all the axioms of G3cp are "NM-G3cp-fit."

Definition 2 *A NM-G3cp-fit base consequence relation is a base consequence relation that includes all the atomic instances of* $\Gamma, p\mid\!\sim_0 p, \Delta$ *and* $\bot, \Gamma\mid\!\sim_0 \Delta$ *(and possibly more sequents).*

---

[10]This is similar to what is done in work on atomic systems (Piecha & Schroeder-Heister, 2016; Sandqvist, 2015). In this literature, it is common to allow higher-order rules and to enforce so-called "definitional reflection." I do neither of these two things here.

Shopper's Guide: Nonmonotonic Logic

Axioms of NM-G3cp

If $\Gamma \mathrel{\vert\!\sim}_0 \Delta$, then $\Gamma \mathrel{\vert\!\sim} \Delta$ is an axiom.

Rules of NM-G3cp (which are identical to those of G3cp)

$$\frac{\Gamma \mathrel{\vert\!\sim} A, \Delta \quad \Gamma, B \mathrel{\vert\!\sim} \Delta}{\Gamma, A \to B \mathrel{\vert\!\sim} \Delta} \text{ Lc} \qquad \frac{\Gamma, A \mathrel{\vert\!\sim} B, \Delta}{\Gamma \mathrel{\vert\!\sim} A \to B, \Delta} \text{ Rc}$$

$$\frac{\Gamma, A, B \mathrel{\vert\!\sim} \Delta}{\Gamma, A \& B \mathrel{\vert\!\sim} \Delta} \text{ L\&} \qquad \frac{\Gamma \mathrel{\vert\!\sim} A, \Delta \quad \Gamma \mathrel{\vert\!\sim} B, \Delta}{\Gamma \mathrel{\vert\!\sim} A \& B, \Delta} \text{ R\&}$$

$$\frac{\Gamma, A \mathrel{\vert\!\sim} \Delta \quad \Gamma, B \mathrel{\vert\!\sim} \Delta}{\Gamma, A \vee B \mathrel{\vert\!\sim} \Delta} \text{ Lv} \qquad \frac{\Gamma \mathrel{\vert\!\sim} A, B, \Delta}{\Gamma \mathrel{\vert\!\sim} A \vee B, \Delta} \text{ Rv}$$

Figure 2: NM-G3cp

If we want to construct a particular nonmonotonic logic, we have to choose a particular NM-G3cp-fit base relation. We should choose the base relation that determines, to the extent that consequence relations do that, the meanings for the atomic sentences that we want them to have. We can now define the nonmonotonic logic NM-G3cp.

Definition 3 NM-G3cp: *A logic belongs to the family NM-G3cp just in case its consequence relation, $\mathrel{\vert\!\sim}$, is the smallest set of sequents that closes an NM-G3cp-fit base consequence relation under the rules of NM-G3cp.*

Figure 2 gives a sequent calculus for NM-G3cp. NM-G3cp has a couple of nice properties. First, it is supra-classical. Second, NM-G3cp extends NM-G3cp-fit base relations conservatively. To see this, notice that all the rules of NM-G3cp conclude sequents that contain logical constants. Third, like in G3cp, all the rules are invertible.

Proposition 1 *All the rules of NM-G3cp are invertible; i.e., if the root sequent of a rule application holds, then so do the top sequents.*

Proof. By induction on proof-height. I will do just one subcase of L&, namely the one where the root comes by LC. Suppose that $\Gamma, A\&B, C \to D\hspace{-0.5em}\mid\hspace{-0.3em}\sim \Delta$ comes by LC. The premises are $\Gamma, A\&B, \hspace{-0.3em}\mid\hspace{-0.3em}\sim C, \Delta$ and $\Gamma, A\&B, D\hspace{-0.5em}\mid\hspace{-0.3em}\sim \Delta$. By our induction hypothesis, this means that we can derive $\Gamma, A, B, \hspace{-0.3em}\mid\hspace{-0.3em}\sim C, \Delta$ and $\Gamma, A, B, D\hspace{-0.5em}\mid\hspace{-0.3em}\sim \Delta$. By LC, we get $\Gamma, A, B, C \to D\hspace{-0.5em}\mid\hspace{-0.3em}\sim \Delta$. The other cases are analogous. □

The invertibility of the rules implies that if we have an effective procedure for checking whether a set of atomic sequents is in our base relation, then the NM-G3cp extension allows for effective root-first proof search.

A further attractive consequence of the invertibility of the G3cp rules is that the connectives satisfy DDT, PF, and FU. That is good news for inferentialists who think that DDT, PF and FU are constitutive of the meanings of the conditional, disjunction, and conjunction respectively.

NM-G3cp is of the type *Rej(Mixed-Cut,CT,CM,DI)*. After all, it is supraclassical (so CO and PEM hold) and it obeys DDT, FU, and PF. Hence, Cautious Monotonicity, Cumulative Transitivity and Distribution all fail in NM-G3cp.

5.3 Making LR nonmonotonic

We have seen in the previous subsection that NM-G3cp obeys neither Mixed-Cut nor CT. That will certainly raise some eyebrows. Although nontransitive logics are becoming more common (Cobreros, Egré, Ripley, & van Rooij, 2012; Ripley, 2013), some people will insist on Cut. In this section, I show how we can construct a nonmonotonic logic whose proof-theory is as elegant as that of NM-G3cp while it also satisfies DDT and Cut. In order to do so, we must turn to relevance logic, as is obvious from Figure 1 because we must reject CO in order to jointly satisfy DDT and CT.

It is well known that the most prominent relevance logics E and R don't have simple sequent calculus formulations because they obey distribution. The distribution-free relevance logic LR, however, has an elegant sequent calculus formulation (Bimbó, 2015). So let's follow the strategy we used to formulate NM-G3cp in order to turn LR into the nonmonotonic logic NM-LR.

The axioms of LR are all the instances of RE (i.e. $A\mid\hspace{-0.3em}\sim A$). For our logic to be as strong as LR, we must ensure that our logic proves all these instances. We can do so by, first, requiring that all base relations include all atomic instances of RE and, second, ensuring that our rules allow us to

Shopper's Guide: Nonmonotonic Logic

Axioms of NM-LR

If $\Gamma \mathrel{\mid\!\sim}_0 \Delta$, then $\Gamma \mathrel{\mid\!\sim} \Delta$ is an axiom.

Rules of NM-LR (which are identical to those of LR)

$$\frac{\Gamma \mathrel{\mid\!\sim} A, \Delta \quad \Theta, B \mathrel{\mid\!\sim} \Lambda}{\Gamma, \Theta, A \to B \mathrel{\mid\!\sim} \Delta, \Lambda} \text{ LC} \qquad \frac{\Gamma, A \mathrel{\mid\!\sim} B, \Delta}{\Gamma \mathrel{\mid\!\sim} A \to B, \Delta} \text{ RC}$$

$$\frac{\Gamma \mathrel{\mid\!\sim} A, \Delta}{\Gamma, \neg A \mathrel{\mid\!\sim} \Delta} \text{ LN} \qquad \frac{\Gamma, A \mathrel{\mid\!\sim} \Delta}{\Gamma \mathrel{\mid\!\sim} \neg A, \Delta} \text{ RN}$$

$$\frac{\Gamma, A_l \mathrel{\mid\!\sim} \Delta}{\Gamma, A_1 \& A_2 \mathrel{\mid\!\sim} \Delta} \text{ L\& } (l=1 \text{ or } 2) \qquad \frac{\Gamma \mathrel{\mid\!\sim} A, \Delta \quad \Gamma \mathrel{\mid\!\sim} B, \Delta}{\Gamma \mathrel{\mid\!\sim} A \& B, \Delta} \text{ R\&}$$

$$\frac{\Gamma, A \mathrel{\mid\!\sim} \Delta \quad \Gamma, B \mathrel{\mid\!\sim} \Delta}{\Gamma, A \vee B \mathrel{\mid\!\sim} \Delta} \text{ Lv} \qquad \frac{\Gamma \mathrel{\mid\!\sim} A_l, \Delta}{\Gamma \mathrel{\mid\!\sim} A_1 \vee A_2, \Delta} \text{ Rv } (l=1 \text{ or } 2)$$

Figure 3: NM-LR

derive all the logically complex instances of RE. With a view to the first point, we define NM-LR-fit base relations.

Definition 4 *A NM-LR-fit base consequence relation is a base consequence relation,* $\mathrel{\mid\!\sim}_0$, *that includes all atomic instances of RE (i.e.* $p \mathrel{\mid\!\sim}_0 p$).

We can now simply use the sequent rules of LR (with the exception that we ignore the rules for contraction because we are working with sets). After all, the rules of LR preserve Reflexivity. The resulting system, NM-LR, is set out in Figure 3.

NM-LR includes LR because we get all the axioms of LR (which can easily be shown by induction on complexity). By inspecting the rules, we can see that NM-LR defines a conservative extension. For, as in NM-G3cp, each rule concludes a sequent with a logical connective. As desired, Cut is admissible in NM-LR if the base relation we choose satisfies Cut.

Proposition 2 *Mixed-Cut is admissible in any NM-LR extension of a NM-LR-fit base relation that satisfies Mixed-Cut.*

Proof. The known Cut-elimination proof for LR, which works basically like Gentzen's original proof for LK (Bimbó, 2015), shows that Mixed-Cut can be pushed up in proof-trees of NM-LR. Hence, it suffices to note that the axioms are closed under Mixed-Cut. □

Notice that NM-LR gives us relevance logics that are ampliative (if the base relation is ampliative). As intimated in Section 2, nonmonotonic logics in the tradition of default logic and KLM want to have more consequences than classical logic, whereas relevance logicians want to have fewer consequences than classical logic. The consequence relations of NM-LR can do both of these things at the same time. We can add risky inferences, and we can do this without getting, e.g., paradoxes of material implication.

6 Conclusion

I have described an exhaustive menu of options from which you can choose a nonmonotonic logic. All nonmonotonic logics fall into at least one of the types charted in Figure 1. Along the way, I have presented a sequent calculus approach to nonmonotonic logic, and I have given two examples of logics that this approach yields, NM-G3cp and NM-LR. I have argued that these logics have some features that are attractive for inferentialists.

References

Anderson, A. R., & Belnap, N. D. (1975). *Entailment: The Logic of Relevance and Necessity*. Princeton: Princeton University Press.

Arieli, O., & Avron, A. (2000). General patterns for nonmonotonic reasoning: from basic entailments to plausible relations. *Logic Journal of the IGPL*, *8*(2), 119–148.

Batens, D. (2007). A universal logic approach to adaptive logics. *Logica Universalis*, *1*(1), 221–242.

Bimbó, K. (2015). *Proof theory: Sequent calculi and related formalisms*. Boca Raton: CRC Press.

Cobreros, P., Egré, P., Ripley, D., & van Rooij, R. (2012). Tolerant, classical, strict. *Journal of Philosophical Logic*, *41*(2), 347–385.

Dung, P. M. (1995). On the acceptability of arguments and its fundamental role in nonmonotonic reasoning, logic programming and n-person games. *Artificial Intelligence, 77*, 321–358.

Dunn, M., & Restall, G. (2002). Relevance logic. In D. Gabbay & F. Guenthner (Eds.), *Handbook of Philosophical Logic* (Vol. 6, pp. 1–128). Kluwer Academic Publishers.

Hlobil, U. (2016). A nonmonotonic sequent calculus for inferentialist expressivists. In P. Arazim & M. Dančák (Eds.), *The Logica Yearbook 2015* (pp. 87–105). College Publications.

Kraus, S., Lehmann, D., & Magidor, M. (1990). Nonmonotonic reasoning, preferential models and cumulative logics. *Artificial Intelligence, 44*, 167-207.

Makinson, D. (2003). Bridges between classical and nonmonotonic logic. *Logic Journal of the IGPL, 11*(1), 69–96.

Piecha, T., & Schroeder-Heister, P. (2016). Atomic systems in proof-theoretic semantics: Two approaches. In J. Redmond, O. Pombo Martins, & Á. Nepomuceno Fernández (Eds.), *Epistemology, Knowledge and the Impact of Interaction* (pp. 47–62). Cham: Springer.

Reiter, R. (1980). A logic for default reasoning. *Artificial Intelligence, 13*(1–2), 81–132. (Special Issue on Non-Monotonic Logic)

Ripley, D. (2013). Paradoxes and failures of Cut. *Australasian Journal of Philosophy, 91*(1), 139–164.

Sandqvist, T. (2015). Base-extension semantics for intuitionistic sentential logic. *Logic Journal of the IGPL, 23*(5), 719–731.

Strasser, C., & Antonelli, G. A. (2016). *Non-monotonic Logic*. The Stanford Encyclopedia of Philosophy (Winter 2016 Edition), Edward N. Zalta (ed.),. Retrieved from https://plato.stanford.edu/ archives/ win2016/ entries/ logic-nonmonotonic/

Troelstra, A. S., & Schwichtenberg, H. (2000). *Basic proof theory* (2nd ed.). Cambridge University Press.

Ulf Hlobil
Concordia University
Canada
E-mail: ulf.hlobil@concordia.ca

Towards an Operational View of Purity

REINHARD KAHLE AND GABRIELE PULCINI[1]

Abstract: A proof is regarded as *pure* in case the technical machinery it deploys to prove a certain theorem does not outstrip the mathematical content of the theorem itself. In this paper, we consider three different proofs of Euclid's theorem affirming the infinitude of prime numbers and we show how, in the light of this specific case study, some of the definitions of purity provided in the contemporary literature prove not completely satisfactory. In response, we sketch the lines of a new approach to purity based on the notion of *operational content* of a certain theorem or proof. Operational purity is here ultimately intended as a way to refine Arana and Detlefsen's notion of 'topical purity'

Keywords: Purity of methods, Infinitude of primes, Hilbert's 24th problem, Mathematical practice

1 Introduction

Purity of methods is a central topic in both logic (especially proof theory) and the philosophy of mathematics. In particular, the quest for purity is related to two key questions for research, one logical and the other epistemological: *Does any theorem admit of a pure proof? What is the actual mathematical gain in providing a pure proof?* In a recent paper on the topic, Andrew Arana attaches to the adjective 'pure' the following meaning: "a proof of a theorem is 'pure' if it draws *only* on what is 'close' or 'intrinsic' to that theorem" (Arana, 2017). This definition can be restated by saying that a proof $\pi : T$ (read: π is a proof of the theorem T) is pure in case the mathematical machinery deployed in π does not transcend the content of T. Put in these terms, the meaning of 'purity' is reduced to the meanings of the

---

[1] We are thankful to Andrew Arana and Marco Panza for their helpful comments and suggestions on earlier versions of this paper. Moreover, we thankfully acknowledge the support from the Portuguese Science Foundation, FCT, through the project "Hilbert's 24th Problem" PTDC/MHCFIL 2583/2014 and the "Centro de Matemática e Aplicações" UID/MAT/00297/2013.

two words 'content' and 'transcend'. Which meanings should be properly attached to these words in order to satisfactorily take account of the basic intuitions coming from mathematical practice is our main concern here.

In spite of the fact that a clear (meta)mathematical definition of the notion of purity — even for a single branch of mathematics! — is still lacking, the history of mathematics offers a wide range of very well-known examples of pure and impure proofs. Take for instance Euclid's theorem affirming the infinitude of primes. On one hand, Euclid's original proof as it appear in the *Elements* (Book IX, Prop. 20) can be considered as perfectly pure. On the other hand, Fürstenberg's topological proof is usually regarded as a sample of impurity, since it resorts to topological considerations which are clearly extraneous to the restricted scope of elementary number theory (Fürstenberg, 1955). Another very well-known example is Wiles' proof of Fermat's Last Theorem whose wild impurity is due to the imbalance between the disarming simplicity of the property stated by the theorem and the overwhelming richness of the technical machinery deployed to devise its proof (elliptic curves, Galois connections, etc.) (McLarty, 2010). To provide a last example of impurity directly coming from the field of mathematical logic (and, as far as we know, never mentioned in the literature on the topic), let us mention semantic proofs of cut-elimination such as, for instance, Okada's lemma for proving cut-redundancy in linear logic (Girard, 1987a; Okada, 1999).

Actually, notions like 'purity' and 'impurity', as well as 'simplicity' and 'explanation', are in general very hard to grasp in their full generality, especially when one tries to proceed by abstraction from a set of concrete examples. As a matter of fact, there are two ways to face this kind of problems: by considering ordinary or formalized mathematics. At first sight, the latter option might seem to have more legs. However, if we take a closer look, this first impression turns out to be quite ill-founded. In proof theory the notion of purity has a natural counterpart in the notion of *analyticity*. A proof $\pi : T$ — developed in a suitable sequent system — is said to be analytic if each one of the formulas displayed in π is a subformula of its conclusion T. In other words, if T is proved by means of an analytic proof π, then any formula appearing in π will be T or a simpler component of T. Thus, if T claims a property in elementary number theory, merely referring to natural numbers, and its proof π is analytic, then π cannot mention any 'extraneous' elements, i.e., any element which is not already contained in T. As is well-known, cut-elimination implies the subformula property, i.e., the fact that any theorem admits an analytic proof (Gentzen, 1938, 1969).

Operational Purity

Unfortunately, the formal grasp of the notion of purity through the notion of analyticity turns out to be defective in three main respects. First, first-order theories obtained by enriching first-order logic by a cluster of proper (i.e., non-logical) axioms do not allow for cut-elimination (Gentzen, 1938, 1969; Girard, 1987b) or, if they do, cut-elimination does not guarantee the subformula property (Negri & von Plato, 2014; Piazza & Pulcini, 2016). The second problem is that the subformula property seems to be something too rigid to be taken as a formal counterpart of the notion of purity to the extent that not 'occurring' in T does not necessarily mean to be an impure element. For instance, many relevant properties about integers are proved in modular arithmetic, though many of them do not mention the notions of 'congruence' or 'modulus'; nonetheless, we cannot see the reason why they should be catalogued as impure. Finally, the third difficulty is about the fact that the implementation of cut-elimination algorithms might make the size of proofs explode (Boolos, 1984). This is to say that, even if analyticity would be an actually achievable goal, the resulting proofs might prove completely intractable objects.

So here is the dilemma about purity as it emerges from contemporary literature on the topic. On one hand, if we decide to refer to formalized mathematics, we unavoidably run into the just-mentioned logical difficulties and limitations (Arana, 2008). On other hand, definitions designed to apply to ordinary theorems and proofs seem to be deemed to indulge on a sort of intrinsic vagueness which, in the worse cases, results in psychologism. This paper should be regarded as a first effort in resolving this dilemma by providing a mathematical definition able to account of purity as it stems from ordinary mathematics.

Summary. In the next section we take the infinitude of prime numbers (henceforth abridged with IP) as a case study. In particular, we review three different proofs of the theorem: Euclid's elementary proof, Fürstenberg's topological proof and, finally, Mercer's topology-free rendition of Fürstenberg's proof. Subsequently, in Section 3, we survey the notion of *topical purity*, recently introduced by Andrew Arana and Michael Detlefsen (Arana, 2017; Arana & Detlefsen, 2011). In Section 4, we sketch an alternative approach to purity based on the notion of *operational content* of a theorem/proof. Finally, in Section 5, we draw some conclusions and indicate the lines of further research.

Reinhard Kahle and Gabriele Pulcini

2 Three proofs of the infinitude of primes

In this section, we consider three different proofs of IP. The first of the list is Euclid's original argument (*Elements*, Book IX, Prop. 20), which is a crystal-clear example of purity. Then we jump in time to Fürstenberg's proof (Fürstenberg, 1955). This proof of IP is worth mentioning because it is in general considered as impure to the extent that it resorts to topological considerations, extraneous to the scope of elementary number theory. The final proof we propose was given in more recent years by Mercer and it is presented as a topology-free version of Fürstenberg's construction (Mercer, 2009).[2]

2.1 Euclid's elementary proof

We report here below a modern rendering of Euclid's original argument as it is reported, for instance, in (Aigner & Ziegler, 2010) and (Arana & Detlefsen, 2011).

Proof. Assume there are exactly n prime numbers, i.e., $\mathbb{P} = \{p_1, \ldots, p_n\}$. Then consider $k = p_1 \cdot p_2 \cdots \cdot p_n + 1$ and let p be one of its prime divisors. If p were in $\mathbb{P}$, then we would have, simultaneously, $p \mid p_1 \cdot p_2 \cdots \cdot p_n$ and $p \mid p_1 \cdot p_2 \cdots \cdot p_n + 1$, hence $k = 1$. But, since there is at least one prime number, $k > 2$. □

This argument can be easily turned into a proof by induction which better emphasizes its constructive nature. Actually: (i) given any finite set of primes $\{p_1, \ldots, p_n\}$, Euclid's argument guarantees the presence of a prime p not in $\{p_1, \ldots, p_n\}$ comprised between 1 and $p_1 \cdot p_2 \cdots \cdot p_n + 1$, and (ii) the property of being a prime number is primitive recursive and so effectively checkable.

Proof (inductive version). We show that, if the set of primes $\mathbb{P}$ has a finite subset $\mathcal{P}_n$ of cardinality n, then it also has a subset $\mathcal{P}_{n+1}$ of cardinality $n+1$. (*Base*) $\mathcal{P}_1 = \{2\} \subseteq \mathbb{P}$. (*Step*) Let $\mathcal{P}_n = \{p_1, \ldots, p_n\}$. Then, consider $k = p_1 \cdot p_2 \cdots \cdot p_n + 1$ and any prime divisor p of k. If p were in $\mathcal{P}_i$, then we would have, simultaneously, $p \mid p_1 \cdot p_2 \cdots \cdot p_n$ and $p \mid p_1 \cdot p_2 \cdots \cdot p_n + 1$, hence $k = 1$. Since $n > 1$, we have that $k > 2$, and so $p \notin \mathcal{P}_n$. Hence, $\mathcal{P}_{n+1} = \mathcal{P}_n \cup \{p\}$ is a subset of $\mathbb{P}$ having cardinality $n + 1$.[3] □

---

[2] A similar topology-free reduction of Fürstenberg's proof can be also found in (Cass & Wildenberg, 2003). Here, we prefer to review Mercer's version because it proves somewhat step by step 'closer' to the original topological argument.

[3] The reader can find a similar inductive version of Euclid's proof in (Arana, 2014).

Operational Purity

2.2 Fürstenberg's topological proof

For $m, r \in \mathbb{Z}$ and $m \geq 1$ consider the two-way infinite arithmetic progression $\mathcal{N}_{m,r} = \{xm + r \mid x \in \mathbb{Z}\}$.

Example 1 $\mathcal{N}_{3,2} = \{\ldots, -4, -1, 2, 5, 8, \ldots\}$.

The set $\mathcal{B} = \{\mathcal{N}_{m,r} \mid m \geq 1\}$ (of subsets of $\mathbb{Z}$) constitutes a *basis* generating a topology on $\mathbb{Z}$, that is:

(i) for any $z \in \mathbb{Z}$, there are $m, r \in \mathbb{Z}$ such that $z \in \mathcal{N}_{m,r}$, i.e., $\bigcup \mathcal{B} = \mathbb{Z}$;

(ii) if $z \in \mathcal{N}_{m_1, r_1} \cap \mathcal{N}_{m_2, r_2}$, then there are $m, r \in \mathbb{Z}$ such that $z \in \mathcal{N}_{m,r} \subset \mathcal{N}_{m_1, r_1} \cap \mathcal{N}_{m_2, r_2}$.

For a detailed proof of these two points the reader can refer to (Arana & Detlefsen, 2011).

Definition 1 (open and closed sets) *A set $S \subseteq \mathbb{Z}$ is said to be open w.r.t. the topology generated by a basis $\mathcal{B}$ if $S = \varnothing$ or, for any $s \in S$, there is a basis element $b \in \mathcal{B}$, such that $s \in b \subseteq S$. A set is closed in case its complement is open.*

Proposition 1 *(i) Any open non-empty set is infinite. (ii) A finite union of closed sets is closed.*

Lemma 1 *Any subset $\mathcal{N}_{m,r}$ of $\mathbb{Z}$ is simultaneously open and closed when considered w.r.t. the topology generated by the basis $\mathcal{B}$.*

Proof. Openness comes straightforwardly from Definition 1 by the fact that any $\mathcal{N}_{m,r}$ is infinite. In order to prove that any element of $\mathcal{B}$ is also closed, write the set $\mathcal{N}_{m,r}$ as follows:

$$\mathcal{N}_{m,r} = \mathbb{Z} - \bigcup_{1 \leq i \leq m-1} \mathcal{N}_{m,r+i}. \qquad (1)$$

Clearly, the union of open sets is still open. By (1), $\mathcal{N}_{m,r}$ proves the complement of an open set, thence it is closed as well. □

Example 2 *Consider the arithmetic progression $\mathcal{N}_{3,2}$ displayed in the previous example. It can be written as*

$$\mathcal{N}_{3,2} = \mathbb{Z} - \bigcup_{1 \leq i \leq 2} \mathcal{N}_{3,2+i} = \mathbb{Z} - (\mathcal{N}_{3,3} \cup \mathcal{N}_{3,4}). \qquad (2)$$

In other words, $\mathcal{N}_{3,2}$ *is the set of the integers whose remainder is 2 when divided by 3. Since* $\mathcal{N}_{3,3} = \mathcal{N}_{3,0}$ *(i.e., the set of the multiples of 3) and* $\mathcal{N}_{3,4} = \mathcal{N}_{3,1}$ *(i.e., the set of all the integers whose remainder is 1 when divided by 3), we can clearly express* $\mathcal{N}_{3,2}$ *as* $\mathbb{Z} - (\mathcal{N}_{3,3} \cup \mathcal{N}_{3,4})$.

Theorem *There are infinitely many primes.*

Proof. Observe that, for any integer $z \notin \{-1, 1\}$: (i) there is a prime p such that $p \mid z$, and (ii) $z \in \mathcal{N}_{p,0}$. Now, assume that the set $\mathbb{P}$ were finite. Thence, we can write:

$$\mathbb{Z} - \{-1, 1\} = \bigcup_{p \in \mathbb{P}} \mathcal{N}_{p,0}. \tag{3}$$

Since $\bigcup_{p \in \mathbb{P}} \mathcal{N}_{p,0}$ comes as the finite union of closed sets, by Proposition $1(ii)$, it is itself closed and so its complement $\{-1, 1\}$ would be simultaneously open and finite, contradicting Proposition $1(i)$. □

2.3 Is Fürstenberg's proof actually impure?

In 2009, Idris Mercer proposed a topology-free version of Fürstenberg's proof (Mercer, 2009). The proof proceeds as follows.

Let $m, r \in \mathbb{Z}$ with $m \leq 1$ and consider the two-way infinite arithmetic progression $f_{m,r}(x) : \mathbb{Z} \mapsto \mathbb{Z}$ defined as $f_{m,r}(x) = mx + r$.

Remark 1 *The range* RAN$(f_{m,r})$ *of* $f_{m,r}$ *contains exactly the integers congruent to r modulo m (i.e., the integers that leave the same remainder $0 \leq r < m$ when divided by m); more formally:*

$$\text{RAN}(f_{m,r}) = \{z \in \mathbb{Z} \mid z \equiv r \,(mod\, m)\}.$$

This means that, for any $k \in \mathbb{Z}$, RAN$(f_{m,r}) = $ RAN$(f_{m,\, km+r})$.

Example 3 RAN$(f_{7,2}) = $ RAN$(f_{7,9}) = $ RAN$(f_{7,-5}) = $
$= \{\ldots, -12, -5, 2, 9, 16, \ldots\} = \{z \in \mathbb{Z} \mid z \equiv 2 \,(mod\, 7)\}$.

Lemma 2 *Any finite intersection of ranges* RAN$(f_{m_1,r_1}) \cap$ RAN$(f_{m_2,r_2}) \cap \cdots \cap$ RAN(f_{m_n,r_n}) *is either empty or infinite.*

Proof. It is clearly possible to have an empty intersection; take, for instance, RAN$(f_{3,2}) \cup$ RAN$(f_{3,0})$. In case it is nonempty, let $z \in $ RAN$(f_{m_1,r_1}) \cap$ RAN$(f_{m_2,r_2}) \cap \cdots \cap$ RAN(f_{m_n,r_n}), i.e., $z = k_1 m_1 + r_1 = k_2 m_2 + r_2 = \cdots = k_n m_n + r_n$. If we take $z' = z + m$ where m is any common multiple of $m_1, \ldots, m_n$, we have that z' also belongs to the intersection. □

Operational Purity

Lemma 3 *Any finite intersection of finite unions of sets can be also expressed as a finite union of finite intersections of the same sets.*

Proof. Easy, by using the fact that intersection distributes over union. □

Theorem *There are infinitely many primes.*

Proof. Let's abbreviate by $\text{NM}(m)$ the finite union of sets:

$$\text{RAN}(f_{m,1}) \cup \text{RAN}(f_{m,2}) \cup \cdots \cup \text{RAN}(f_{m,m-1}) \tag{4}$$

which gathers all the integers *not* divisible by m. The proof proceeds by *reductio*. Assume there are only k prime numbers $\{p_1, p_2, \ldots, p_k\}$. This would imply:

$$\text{NM}(p_1) \cap \text{NM}(p_2) \cap \cdots \cap \text{NM}(p_n) = \{1, -1\}. \tag{5}$$

Now, the set displayed on the left of the equality symbol is a finite intersection of finite unions and so, by Lemma 3, it can be rewritten as a finite union of finite intersections. By Lemma 2, each one of the intersections appearing in the union is either empty or infinite, thence their union is either empty or infinite, which is a contradiction. □

Mercer's proof is particularly interesting for our present purposes because it reveals the possibility of having cases of apparent impurity, i.e., cases in which a proof $\pi : T$ relies on concepts that are conceptually far away from the mathematical content of the theorem T, *but the resort to such extraneous notions is nothing else but an avoidable roundabout*. Fürstenberg's argument is developed around the topological notions of 'basis for a topology', 'open set', and 'closed set'. Nonetheless, these concepts are completely avoidable. In particular, the very fact that the set $\mathcal{B}$ constitutes the basis for a topology on $\mathbb{Z}$ never comes into play, because no topology on $\mathbb{Z}$ ever comes into play. Similarly, the topological notions of openness and closedness can be replaced by easy results involving the basic set-theoretic operations of union and intersection.

This is to say, that the act of classifying a certain proof $\pi : T$ is not just a matter of confronting the 'content' of π with the rough 'content' of T. Actually, one should be able to rule out the possibility of rewriting the proof π into a more elementary version π' whose 'content' does not outstrip the 'content' of T any more.

3 Topical purity

The notion of *topical purity* as it has been defined by Andrew Arana and Michael Detlefsen in (Arana, 2014; Arana & Detlefsen, 2011) is one of the most accurate characterizations of purity currently available in the literature. In (Arana, 2014), the author writes:

> [...] a solution to a problem is "topically pure" if it draws only on what is "contained" in (the content of) that problem, where what is "contained" in a problem is what grounds its understanding and is what we call that problem's "topic". If a solution to a problem draws on something extrinsic to that problem's topic, then it is topically impure. (Arana, 2014, pag. 316)

Here, the 'topic' of a problem has to be thought of as an epistemic notion which gathers all the technical aspects (definitions, axioms, proofs, etc.) that determine the understanding of a particular problem. For instance, according to this approach, the 'topic' of IP would be something formed by the set of axioms needed to understand the linearly ordered structure of positive integers — i.e., the cluster of Peano-Dedekind first-order axioms — together with the definition of being a prime number (through the notion of divisibility).

What distinguishes purity from impurity is the idea that pure solutions to a certain problem are, so to speak, *stable under retraction of their argumentative components*. In order to further clarify this point, Arana continues:

> The heart of the account in (Arana & Detlefsen, 2011) of the epistemic value of topical purity is the following counterfactual: if a component of a topically pure solution to a problem were retracted by an investigator, then that investigator's understanding of that problem would change. (Arana, 2014, pag. 317)

In this paper we uncritically accept the fact that Fürstenberg's and Mercer's proofs actually express the same mathematical argument. Of course, the general constraints under which this kind of 'identity' can be established are far from being clear and well stipulated. However, if we take this identity as unproblematic, it turns out that Fürstenberg's proof has to be classified as topically pure since, once that the topological superstructure has been removed, it still provides a complete solution to the IP problem.

4 Operational purity

4.1 Operational content and operational purity

In this section, we propose a new definition of purity which, we believe, allows for a better grasp of what Arana and Detlefsen call the 'topic' of a theorem/proof. Actually, a good definition of purity should be expected to meet, as much as possible, the following two *desiderata*: (i) avoiding any reference to a knowing subject (as the "investigator's understanding"), and (ii) consider proofs modulo useless roundabouts, i.e., technical or terminological *detours* which prove inessential with respect to the actual argumentative structure of the proof.

Here, the guiding idea consists in putting mathematical operations to the fore and leave ontology on the background as a derived notion. This position is mainly inspired by Waismann's philosophy of mathematics (Waismann, 2003) and in some ways echoes Lagrange's methodological ideal as it has been retraced in (Ferraro & Panza, 2012), according to which, a proof should be classified as pure or impure only relatively to an ideal mathematical background. In particular, what he had in mind is a sort of operational closure under the algebra of polynomials.

Definition 2 (operational content, ontology) *Given a theorem T (resp. a proof π), its operational content $\mathfrak{C}(T)$ (resp. $\mathfrak{C}(\pi)$) is given by the set of* mathematical operations *mentioned in T (resp. π). Given an operational content c, its ontology $\mathfrak{O}(c)$ is defined as the* smallest numerical domain *($\mathbb{N}, \mathbb{Z}, \mathbb{Q}, \ldots$) closed under the operations in c.*

Definition 3 *Given any two operational contents c_1 and c_2, we write $c_1 \preceq c_2$ (resp. $c_1 \prec c_2$) in case $\mathfrak{O}(c_1) \subseteq \mathfrak{O}(c_2)$ (resp. $\mathfrak{O}(c_1) \subset \mathfrak{O}(c_2)$).*

Example 4 *If $c = \{+, \cdot\}$ and $d = \{+, -\}$, then $\mathfrak{O}(c) = \mathbb{N}$ and $\mathfrak{O}(d) = \mathbb{Z}$, thence $c \prec d$.*

Definition 4 (purity, degrees of purity) *We say that $\pi : T$ is pure in case $\mathfrak{C}(\pi) \preceq \mathfrak{C}(T)$. Given two proofs $\varphi, \psi : T$ we say that φ is purer than ψ when $\varphi \prec \psi$.*

Example 5 (operational content of IP) *The definition of prime number relies on the division operation, therefore $\mathfrak{C}(\mathsf{IP}) = \{/\}$. Thence, according to Definition 2, $\mathfrak{O}(\mathfrak{C}(\mathsf{IP})) = \mathbb{Q}$.*

Reinhard Kahle and Gabriele Pulcini

4.2 Operational purity at work

4.2.1 Number theory

Our operational approach has the immediate effect of expanding from $\mathbb{N}$ to $\mathbb{Q}$ the ontology underlying classical arithmetic. In spite of the fact that elementary number theory ultimately deals with properties of natural numbers, the smallest numerical domain closed under the four basic arithmetical operations $\{+, \cdot, -, /\}$ is $\mathbb{Q}$ rather than $\mathbb{N}$.

Concerning Fürstenberg's proof $\psi : \mathsf{IP}$, let us now overlook all the topological apparatus and focus on ψ's operational content. After a careful parsing of the proof, it is easy to acknowledge that $\mathfrak{C}(\psi) = \{+, \cdot, -, /, \cap, \cup\}$. In particular, the definitions of 'open' and 'closed' set are provided simply by mentioning the two set-theoretic operations $\cap$ and $\cup$. Since the inclusion of union and intersection in any operational content does not affect the underlying ontology, we get $\mathfrak{O}(\mathfrak{C}(\psi)) = \mathbb{Q}$ and so $\mathfrak{C}(\psi) \preceq \mathfrak{C}(\mathsf{IP})$, namely Fürstenberg's proofs is recognized as (operationally) pure.

4.2.2 Complex numbers and the real plane

Complex Analysis turned out to be an extremely useful tool to prove statements which are formulated in Real Analysis. The intuitive reason is clear: considering the real line as a cut in the complex plane, functions defined over this plane may provide much more (mathematically accessible) structure than their restriction to the real numbers. The complex numbers, however, seem clearly to be an impure element in a proof of a theorem speaking about real numbers only. It is tempting to see in the use of complex numbers, notably $\sqrt{-1}$, a violation of ontological purity.

Taking the isomorphism of $(\mathbb{C}, +, \cdot)$ and $(\mathbb{R}^2, \oplus, \odot)$, where $\oplus$ is the component-by-component addition of pairs, and $\odot$ is defined by $(a, b) \odot (c, d) = (ac - bd, ad + bc)$, into account, the ontological impurity seems to be overturned. Of course, one is paying the price of an additional "dimension", but it seems to be exaggerated to blame the use of pairs of real numbers as a reason for impurity.

In our operational perspective, the impurity would come from the fact that $\odot$ is not supposed come automatically with the operations defined for $\mathbb{R}$. In formal terms, for a theory T of real analysis, we would say that $\odot \notin \mathfrak{C}(T)$. And, based on the above-mentioned isomorphism, $\mathfrak{O}(\oplus, \odot)$ is not just $\mathbb{R}^2$ but, indeed, $\mathbb{C}$.

5 Purity and Hibert's 24th problem

Purity of Method has a long history in Mathematics, and it was prominently addressed by David Hilbert in the final paragraph of his seminal book *Grundlagen der Geometrie*:

> In modern mathematics such criticism is raised very often, where the aim is to preserve the purity of method, i.e. to prove theorems if possible using means that are suggested by the content of the theorem. (Hilbert, 2004, pag. 315-16)

Hilbert's interests in properties of mathematical proofs were very broad. In 2000, a draft note of David Hilbert was found in his notebook concerning a possible 24th problem for his Paris lecture in 1900 (Hilbert, 1901); this problem concerns *simplicity of proofs*:

> **H24**: Criteria of simplicity, or proofs of the greatest simplicity of certain proofs. Develop a theory of the method of proof in mathematics in general. Under a given set of conditions there can be but one simplest proof. Quite generally, if there are two proofs for a theorem you must keep going until you have derived each from the other, or until it becomes quite evident what variant conditions (and aids) have been used in the two proofs. Given two routes, it is not right to take either of these two or to look for a third; it is necessary to investigate the area underlying between the two routes. (Thiele, 2003)

When one asks for possible relations of purity and simplicity (in the way, Hilbert addressed the two problems) one may observe that Hilbert explicitly asks for simplicity "under a given set of conditions". In general, proofs with different degrees of purity will change the "set of conditions" (i.e., the axiomatic frameworks in which the proofs are performed) and, in a strict reading, a comparison of such proofs would not fall within the scope of Hilbert's 24th problem. However, when Hilbert asked for "what variant conditions (and aids) have been used in [...] two proofs", such variations may well be classified in terms of purity. If so, the request "to investigate the area underlying between the two routes" could be tied to the question of purity. For such a relation, however, it seems to be reasonable to consider purity and simplicity as complementary. Considering again theorems of real analysis where one may draw on complex analysis (see §4.2.2), a purely ontological perspective may classify proofs which only use real numbers as simple as those which resort to complex numbers (as real numbers are considered to be simpler than complex numbers). From an operational perspective, however, it seems to be reasonable that proofs using complex numbers are, indeed, simpler as the invoked operations can be considered as simple as the operations needed to be defined for a pure proof.

Needless to say, the operational definition of purity we offered here is just a first proposal which calls for further refinement and confrontation with a richer variety of mathematical examples. *In primis*, in the context of the IP problem, operational purity should be tested with proofs which seem to be genuinely impure, such as Euler's argument based on harmonic series (Aigner & Ziegler, 2010; Mancosu & Marion, 2010).

Moreover, we acknowledge that our definition might present some intrinsic limits due to its heavy reliance on the notions of mathematical operation and set-theoretic closure under a given set of operations. These notions both prove clear when considered, for instance, in number theory, but they become somewhat vague in other fields such as geometry. To take an example, in Euclidean geometry, there seems to be no clear conceptual cleavage between constructions specifically pertaining to the plane and constructions which require the third dimension. Thus, our operational definition seems to be of no immediate help in controversial cases such as that of Desargues' theorem in which the third dimension is needed to prove a property of plane geometry (Arana & Mancosu, 2012). In this perspective, analytic geometry could help in clarifying the different kinds of geometric constructions in terms of algebraic operations.

References

Aigner, M., & Ziegler, G. M. (2010). *Proofs from the Book* (Vol. 274). Springer.

Arana, A. (2008). On formally measuring and eliminating extraneous notions in proofs. *Philosophia Mathematica*, *17*(2), 189–207.

Arana, A. (2014). Purity in arithmetic: some formal and informal issues. In G. Link (Ed.), *Formalism and Beyond.* Walter de Gruyter Inc.

Arana, A. (2017). On the alleged simplicity of impure proof. In R. Kossak & P. Ording (Eds.), *Simplicity: Ideals of Practice in Mathematics and the Arts*. Springer International Publishing.

Arana, A., & Detlefsen, M. (2011). Purity of methods. *Philosophers' Imprint, 11*(2).

Arana, A., & Mancosu, P. (2012). On the relationship between plane and solid geometry. *The Review of Symbolic Logic, 5*(2), 294–353.

Boolos, G. (1984). Don't eliminate cut. *Journal of Philosophical Logic, 13*(4), 373–378.

Cass, D., & Wildenberg, G. (2003). Math bite: A novel proof of the infinitude of primes, revisited. *Mathematics Magazine, 76*(3), 203.

Ferraro, G., & Panza, M. (2012). Lagrange's theory of analytical functions and his ideal of purity of method. *Archive for History of Exact Sciences, 66*(2), 95–197.

Fürstenberg, H. (1955). On the infinitude of primes. *American Mathematical Monthly, 62*(5).

Gentzen, G. (1938). Neue Fassung des Widerspruchsfreiheitsbeweises für die reine Zahlentheorie. *Forschungen zur Logik und zur Grundlegung der exakten Wissenschaften, 4*(1), 19–44.

Gentzen, G. (1969). *The Collected Papers of Gerhard Gentzen*. Amsterdam: North-Holland Pub. Co.

Girard, J.-Y. (1987a). Linear logic. *Theor. Comput. Sci., 50*, 1–102.

Girard, J.-Y. (1987b). *Proof Theory and Logical Complexity*. Bibliopolis.

Hilbert, D. (1901). Mathematical problems. *Bull. Amer. Math. Soc., 8*, 437–479.

Hilbert, D. (2004). *Lectures on the Foundations of Geometry, 1891-1902*. Springer.

Mancosu, P., & Marion, M. (2010). Wittgenstein's constructivization of Euler's proof of the infinity of primes. In P. Mancosu (Ed.), *The Adventure of Reason: Interplay Between Philosophy of Mathematics and Mathematical Logic, 1900-1940*. Oxford University Press.

McLarty, C. (2010). What does it take to prove Fermat's last theorem? Grothendieck and the logic of number theory. *Bulletin of Symbolic Logic, 16*(3), 359–377.

Mercer, I. D. (2009). On Fürstenberg's proof of the infinitude of primes. *American Mathematical Monthly, 116*.

Negri, S., & von Plato, J. (2014). *Proof Analysis — A Contribution to Hilbert's Last Problem*. Cambridge University Press.

Okada, M. (1999). Phase semantic cut-elimination and normalization proofs of first- and higher-order linear logic. *Theor. Comput. Sci., 227*(1-2), 333–396.

Piazza, M., & Pulcini, G. (2016). Uniqueness of axiomatic extensions of cut-free classical propositional logic. *Logic Journal of the IGPL*, *24*(5), 708–718.

Thiele, R. (2003). Hilbert's twenty-fourth problem. *American Mathematical Monthly*, *110*(1), 1–24.

Waismann, F. (2003). *Introduction to Mathematical Thinking: The Formation of Concepts in Modern Mathematics*. Dover.

Reinhard Kahle
Universidade Nova de Lisboa
CMA & DM, FCT
Portugal
E-mail: `kahle@mat.uc.pt`

Gabriele Pulcini
Universidade Nova de Lisboa
CMA, FCT
Portugal
E-mail: `g.pulcini@fct.unl.pt`

A Multi-Succedent Sequent Calculus for Logical Expressivists

DAN KAPLAN[1]

Abstract: Expressivism in logic is the view that logical vocabulary plays a primarily expressive role: that is, that logical vocabulary makes perspicuous in the object language structural features of inference and incompatibility (Brandom, 1994, 2008). I present a precise, technical criterion of expressivity for a logic (§2). I next present a logic that meets that criterion (§3). I further explore some interesting features of that logic: first, a representation theorem for capturing other logics (§3.1), and next some novel logical vocabulary for *expressing* structural features of inference (§4).

Keywords: Inferentialism, Logical expressivism, Non-monotonic logic

1 Introduction: some philosophical background

In this paper I present a non-monotonic, multi-succedent sequent calculus that vindicates the ambitions of logical expressivists and semantic inferentialists. Expressivism in logic is the view that logical vocabulary plays a primarily expressive role: that is, that logical vocabulary makes perspicuous *in the object language* structural features of inference and incompatibility (Brandom, 1994, 2008). Brandom cashes this out with the slogan that logical vocabulary allows one to *say* (in the object language) what one was previously only able to *do* (in a pre-logical discursive practice). The result is that logical vocabulary should be understood as "LX", i.e. *(algorithmically) elaborated from* and *explicative of* a pre-logical consequence relation. (Algorithmic) elaboration is a pragmatic constraint: the ability to competently navigate such a pre-logical consequence relation already endows one with the abilities needed to navigate a consequence relation with logical vocabulary. That such vocabulary is explicative means that it must successfully encode structural features of that consequence relation.

---

[1] This work is the fruit of a joint research project. I am indebted to the helpful feedback given by a research group run by Bob Brandom. The central philosophical and technical results I am reporting, however, are my own. A system with similar ambitions (in fact an earlier result in our project) can be seen in (Hlobil, 2016).

Dan Kaplan

Logical expressivism is motivated in turn by two other significant philosophical theses: *semantic* inferentialism—the view that the meaning of *non-logical* vocabulary is determined (at least, if not essentially) by its role in inference—and *logical* inferentialism—the view that the meaning of *logical* vocabulary is also so determined, paradigmatically by introduction and elimination rules (Brandom, 1994, 2008; Peregrin, 2014). It is the former of these two views that distinguishes expressivism from logical inferentialism. Since our ordinary discursive practices potentially include material and non-monotonic inferences, the logical expressivist wishes to understand logical vocabulary as expressive of *those* inferential practices.

Combining these three lines of thought produces some demands on a logical system. A commitment to semantic inferentialism means that (i) our logical systems should include within them material and pre-logical fragments, out of which logical vocabulary is to be *elaborated*. (ii) Such elaboration should in turn *naturally and conservatively* extend such a pre-logical consequence relation. The extension should be *conservative* in the sense that no new material implications are introduced as a result; further, the structural features that our logical vocabulary expresses should likewise be preserved in the logically extended consequence relation. The demand that the extension be *natural* means that specifying the inferential role of logical vocabulary should suffice *on its own* to extend and preserve the structural features in question (e.g. that no further structural rules are required). This demand (if met) justifies the claim that the abilities needed to use non-logical vocabulary *already* endow one with the abilities needed to use logical vocabulary. Finally, (iii) such systems should be capable of expressing *in the object language* those features of consequence that were discarded as a result of (i); they should, therefore, express when an implication is e.g. monotonic or classically valid.

The system I construct meets these demands. In order to show this, I begin by making precise what logical expressivism demands. That is, I argue for a precise criterion against which a logic may be tested (§2). I also argue for two additional criteria intended to make precise the idea that a logic preserves and expresses *structural* features of implication. Following this I construct a system that meets all of these constraints (§3). I start with a material, non-monotonic, multi-succedent base consequence relation over an atomic language: $\mathord{\mid\kern-0.4em\sim}_0 \subseteq \mathcal{P}(\mathcal{L}_0) \times \mathcal{P}(\mathcal{L}_0)$. The language is extended in the standard fashion to include $\{\&, \vee, \neg, \rightarrow\}$. The consequence relation is conservatively extended to $\mathord{\mid\kern-0.4em\sim} \subseteq \mathcal{P}(\mathcal{L}) \times \mathcal{P}(\mathcal{L})$ via familiar sequent rules. The extension is also natural in the sense sketched above. Local

A Multi-Succedent Sequent Calculus for Logical Expressivists

regions of monotonicity and classical validity are each preserved by the logical rules of our sequent calculus without the aid of any structural rules. This means, in the case of classicality, that the stipulation that our base consequence relation contain all atomic classical sequents (i.e. sequents of the form $\Gamma, p \mathrel{\mid\!\sim}_0 p, \Theta$ for $p \in \mathcal{L}_0$, $\Gamma, \Theta \subseteq \mathcal{L}_0$) guarantees that our logically extended consequence relation be supra-classical via the logical rules alone. That monotonicity is *naturally* preserved means that if a base sequent is monotonic with respect to atoms $\forall \Delta_0, \Lambda_0 \subseteq \mathcal{L}_0(\Delta_0, \Gamma_0 \mathrel{\mid\!\sim}_0 \Theta_0, \Lambda_0)$, then that sequent will also be monotonic with respect to logically complex sentences: $\forall \Delta, \Lambda \subseteq \mathcal{L}(\Delta, \Gamma_0 \mathrel{\mid\!\sim} \Theta_0, \Lambda)$. And this result holds for *all* implications in the logically extended consequence relation:

$$\forall \Delta_0, \Lambda_0 \subseteq \mathcal{L}_0(\Delta_0, \Gamma \mathrel{\mid\!\sim} \Theta, \Lambda_0) \Rightarrow \forall \Delta, \Lambda \subseteq \mathcal{L}(\Delta, \Gamma \mathrel{\mid\!\sim} \Theta, \Lambda).$$

In addition, I introduce modal operators to encode these very same structural features in the object language (satisfying (iii) above). That is, I show how we can introduce an operator '$\boxed{M}$' which marks all and only those sequents which obey monotonicity:

$$\forall \Delta, \Lambda \subseteq \mathcal{L}(\Delta, \Gamma \mathrel{\mid\!\sim} A, \Theta, \Lambda) \Leftrightarrow \Gamma \mathrel{\mid\!\sim} \boxed{M} A, \Theta.$$

I also introduce an operator '$\boxed{K}$' which marks all and only classically valid implications:

$$\Gamma \vdash_{LK} A, \Theta \Leftrightarrow \Gamma \mathrel{\mid\!\sim} \boxed{K} A, \Theta.$$

These operators are introduced via simple and straightforward sequent rules and they mark precisely the structural features they purport to mark in virtue of those sequent rules alone. I also explain how the techniques used to develop '$\boxed{M}$' and '$\boxed{K}$' may be generalized to other structural features (§4). It is in virtue of this that I claim we can see the logical vocabulary of my system as truly *expressive of* an underlying material consequence relation.

Finally, I present a representation theorem which allows specification of exactly which implications must be included in the atomic fragment of the consequence relation if we want our extended consequence relation to *include* a potentially arbitrary, logically complex consequence relation (§3.1). If that consequence relation meets several modest syntactic constraints then we may specify *exactly* which base consequence relation will generate *exactly* that logically complex consequence relation. Thus, because the sequent rules I employ *naturally* extend an underlying material and pre-logical consequence relation, we should understand that logical vocabulary as truly algorithmically elaborated from that pre-logical consequence relation. And because my representation theorem allows us to see *exactly* which base implications in a pre-logical consequence relation are responsible for a given implication, we should understand those logically complex sequents as truly *expressive of* that underlying base consequence relation. My system therefore vindicates some core ambitions of logical expressivists.

2 Precisification of "expressivity"

I now seek to make the notion of "expression" more precise. Brandom understands expressivism in terms of what he calls an "LX relation", where a vocabulary B is "LX" of a vocabulary A if it is <u>e</u>laborated from and e<u>x</u>plicative of A. The first criterion (elaboration) has it that if one is able to successfully deploy vocabulary A then one already has the skills necessary to use B. That is, that B may be (algorithmically) elaborated from A. The second criterion (explication) has it that B says something about (makes perspicuous in the object language) what one was doing by using A (minimally that B may encode the implications and incompatibilities of A). Logical vocabulary is said to be "universally LX" meaning that logical vocabulary stands in this relation to *all vocabularies*.

Let us make this relation more precise. First let $\mathcal{L}_0$ be an arbitrary vocabulary devoid of logical symbols (i.e. a set of atomic sentence letters). Let $\mathrel{\mid\!\sim}_0$ be a consequence relation over $\mathcal{L}_0$ (i.e. $\mathrel{\mid\!\sim}_0 \subseteq \mathcal{P}(\mathcal{L}_0)^2$). Note that while I call $\mathrel{\mid\!\sim}_0$ and $\mathrel{\mid\!\sim}$ (below) *consequence* relations I do not yet impose *any* restrictions on them. They should be treated, therefore, simply as two-place relations between sets of sentences. As I discussed in the introduction, there are philosophically rich reasons for wanting a consequence relation that is e.g. non-monotonic or perhaps non-classical. In addition part of the motivation of expressivism is that where such features hold of consequence it is an *expression* of an underlying (material) relation of consequence.[2]

Next let $\mathbb{L}$ be our logic. Our logic consists of a finite set of logical symbols (e.g. $\{\&, \vee, \neg, \rightarrow\}$) and rules for expanding $\mathcal{L}_0$ to $\mathcal{L}$ (our language enriched with those logical symbols) and for expanding $\mathrel{\mid\!\sim}_0$ to $\mathrel{\mid\!\sim}$. Intuitively, we should think of $\mathbb{L}$ as a function from $\mathrel{\mid\!\sim}_0$ to $\mathrel{\mid\!\sim}$. That is, $\mathbb{L} : \mathrel{\mid\!\sim}_0 \mapsto \mathrel{\mid\!\sim}$. Then whether $\mathbb{L}$ is "LX" concerns the relationship between $\mathrel{\mid\!\sim}_0$ and $\mathrel{\mid\!\sim}$ (i.e. the behavior of $\mathcal{L}$ in relation to the behavior of $\mathcal{L}_0$).

---

[2] I am being brief here on the *justification* for treating $\mathrel{\mid\!\sim}_0$ as I do. I primarily wish to stress here—in order to avoid confusion—that $\mathrel{\mid\!\sim}_0$ need not have *any constraints*.

A Multi-Succedent Sequent Calculus for Logical Expressivists

That the logical vocabulary be elaborated fixes a tight relationship from $\mathord{\vert\!\sim}_0$ to $\mathord{\vert\!\sim}$. That is, to get from $\mathord{\vert\!\sim}_0$ to $\mathord{\vert\!\sim}$ should require no more than a specification of the logical vocabulary. That is, given $\mathord{\vert\!\sim}_0$, $\mathord{\vert\!\sim}$ should be uniquely determined: $\mathord{\vert\!\sim}_0 \Rightarrow \mathord{\vert\!\sim}$. In prose, the inferential behavior of $\mathcal{L}$ should be determined by that of $\mathcal{L}_0$ simply by specifying the logical symbols.

That the logical vocabulary be explicative fixes a tight relationship from $\mathord{\vert\!\sim}$ to $\mathord{\vert\!\sim}_0$. Since this requires that the logical vocabulary enable us to *say something* about the underlying pre-logical consequence relation, we should require that it actually do what it purports to do: $\mathord{\vert\!\sim} \Rightarrow \mathord{\vert\!\sim}_0$. In prose, the inferential behavior of $\mathcal{L}$ should *genuinely* say or *express* something about the inferential behavior of $\mathcal{L}_0$. The inferential behavior of $\mathcal{L}$ should therefore fix that of $\mathcal{L}_0$. If $\mathcal{L}$ behaved differently then it would *express* something different about the behavior of $\mathcal{L}_0$. If such expression is to be genuine then the behavior of $\mathcal{L}_0$ (i.e. $\mathord{\vert\!\sim}_0$) would need to be different.

Together these two criteria have it that $\mathord{\vert\!\sim}_0 \Leftrightarrow \mathord{\vert\!\sim}$. The behavior of $\mathcal{L}$ is elaborated out of, but also explicative of the behavior of $\mathcal{L}_0$.

While this criterion has some naive plausibility, it must still be made more precise. In particular, if our logical vocabulary is to be *conservative*, then $\mathord{\vert\!\sim}_0 \subseteq \mathord{\vert\!\sim}$ and so the criterion will hold trivially. We may circumvent this problem by quantifying over possible $\mathord{\vert\!\sim}_0$. This might have already been anticipated since I mentioned that logical vocabulary is to have this relationship *universally* i.e. with respect to arbitrary vocabularies (and thus arbitrary $\mathord{\vert\!\sim}_0$). This gives rise to the following definition:

Definition 1 *Fix a logic* $\mathbb{L}$, *i.e. a function from* $\mathord{\vert\!\sim}_0$ *to* $\mathord{\vert\!\sim}$. *We say that* $\mathbb{L}$ *is* **expressive** *or that* $\mathord{\vert\!\sim}$ **expresses** *a base consequence relation* $\mathord{\vert\!\sim}_0$ *iff*:

$$(\forall \Gamma, \Theta \subseteq \mathcal{L})(\exists \Gamma_1, \Theta_1, \ldots, \Gamma_n, \Theta_n \subseteq \mathcal{L}_0)$$
$$(\forall \mathord{\vert\!\sim}_0 \subseteq \mathcal{P}(\mathcal{L}_0)^2)(\forall \mathord{\vert\!\sim} \subseteq \mathcal{P}(\mathcal{L})^2)(\mathbb{L} : \mathord{\vert\!\sim}_0 \mapsto \mathord{\vert\!\sim}))$$
$$((\Gamma \mathrel{\vert\!\sim} \Theta) \Leftrightarrow (\Gamma_1 \mathrel{\vert\!\sim}_0 \Theta_1 \bigwedge \Gamma_2 \mathrel{\vert\!\sim}_0 \Theta_2 \bigwedge \cdots \bigwedge \Gamma_n \mathrel{\vert\!\sim}_0 \Theta_n)).$$

We also say $\Gamma \mathrel{\vert\!\sim} \Theta$ **expresses** $\Gamma_i \mathrel{\vert\!\sim}_0 \Theta_i$ $(1 \leq i \leq n)$ *(its expressientia).*

This definition says that whenever $\Gamma \mathrel{\vert\!\sim} \Theta$ holds this is in virtue of some set of implications present in the language prior to $\mathbb{L}$. So the logical vocabulary is said to be elaborated if $\Gamma \mathrel{\vert\!\sim} \Theta$ holds whenever those pre-logical implications obtain, and the logical vocabulary is said to be explicative if $\Gamma \mathrel{\vert\!\sim} \Theta$ holds only if those pre-logical implications obtain.

Dan Kaplan

The above should be taken as a precise specification of a minimal constraint on logics to count as "expressive". But one of the central features of expression is the idea that logical vocabulary should be able to make perspicuous in the object language *structural features* of inference. By structural features I have in mind such things as monotonicity, classicality, reflexivity, contraction, etc., where each is understood to be capable of holding both globally (e.g. that $\mathrel{\vert\!\sim}$ is monotonic) as well as locally (e.g. that $\Gamma \mathrel{\vert\!\sim} \Theta$ is monotonic, though $\Delta \mathrel{\vert\!\sim} \Lambda$ may not be). Expressivism says that it is distinctive of logical vocabulary to be able to express such features. This requires that (i) $\mathbb{L}$ be capable of *preserving* structural features and that (ii) $\mathbb{L}$ be capable of *expressing* those very structural features it preserves.

Definition 2 *Let $\mathfrak{Sf}$ be a structural feature. Let $\mathfrak{Sf}(\Gamma \mathrel{\vert\!\sim} \Theta)$ be shorthand for $\Gamma \mathrel{\vert\!\sim} \Theta$ obeys (or is an instance of) $\mathfrak{Sf}$. Next, let $\Gamma \mathrel{\vert\!\sim} \Theta$ be arbitrary with $\Gamma_i \mathrel{\vert\!\sim}_0 \Theta_i$ ($1 \leq i \leq n$) its expressientia (in accordance with Definition 1). We say that a logic $\mathbb{L}$* **preserves** *a structural feature $\mathfrak{Sf}$ iff:*

$$\mathfrak{Sf}(\Gamma \mathrel{\vert\!\sim} \Theta) \Leftrightarrow \left(\mathfrak{Sf}(\Gamma_1 \mathrel{\vert\!\sim}_0 \Theta_1) \bigwedge \cdots \bigwedge \mathfrak{Sf}(\Gamma_n \mathrel{\vert\!\sim}_0 \Theta_n) \right).$$

A structural feature is *preserved* when: an implication obeys that structural feature *iff* all of the implications it expresses also obey that structural feature. Thus, whether an implication obeys a structural feature should be seen as expressing something about the pre-logical implications that that implication expresses: it inherits those features from them and has those features in virtue of those implications alone. Next, I must explain what it means for a particular piece of logical vocabulary to express such structural features.

Definition 3 *Let $\mathfrak{Sf}$ be a structural feature. Suppose some logical operation '$*$' may be used to mark a sequent in some way (with the constraint that $\Gamma^* \mathrel{\vert\!\sim} \Theta^*$ only if $\Gamma \mathrel{\vert\!\sim} \Theta$). Then we say that '$*$' (or $\mathbb{L}$) expresses $\mathfrak{Sf}$ iff there exists a '$*$' in $\mathbb{L}$ such that:*

$$\Gamma^* \mathrel{\vert\!\sim} \Theta^* \Leftrightarrow \mathfrak{Sf}(\Gamma \mathrel{\vert\!\sim} \Theta).$$

This definition of structural feature preservation combines three ideas. (i) That a logic be capable of expressing an underlying base consequence relation, (ii) that it be capable of preserving structural features of that base consequence relation, and finally (iii) that it be able to mark in the object language those very same features that it preserves.

3 A non-monotonic multi-succedent sequent calculus

I now construct a logic $\mathbb{L}$ which I will use to exhibit some of the above ideas. Let us fix $\mathbin{\vert\!\sim}_0$. Let our logic include the symbols $\{\&, \vee, \neg, \rightarrow\}$ and let it expand $\mathcal{L}_0$ to $\mathcal{L}$ in the standard fashion. Then our logic $\mathbb{L}$ is given by the following sequent calculus, where proof trees are introduced by axioms:[3]

Axiom 1: If $\Gamma \mathbin{\vert\!\sim}_0 \Theta$, then $\Gamma \mathbin{\vert\!\sim} \Theta$ may form the base of a proof tree.

$$\frac{\Gamma \mathbin{\vert\!\sim} \Theta, A \quad B, \Gamma \mathbin{\vert\!\sim} \Theta}{A \rightarrow B, \Gamma \mathbin{\vert\!\sim} \Theta} \, \text{L}\!\rightarrow \qquad \frac{A, \Gamma \mathbin{\vert\!\sim} \Theta, B}{\Gamma \mathbin{\vert\!\sim} \Theta, A \rightarrow B} \, \text{R}\!\rightarrow$$

$$\frac{\Gamma, A, B \mathbin{\vert\!\sim} \Theta}{\Gamma, A\&B \mathbin{\vert\!\sim} \Theta} \, \text{L\&} \qquad \frac{\Gamma \mathbin{\vert\!\sim} A, \Theta \quad \Gamma \mathbin{\vert\!\sim} B, \Theta}{\Gamma \mathbin{\vert\!\sim} A\&B, \Theta} \, \text{R\&}$$

$$\frac{A, \Gamma \mathbin{\vert\!\sim} \Theta \quad B, \Gamma \mathbin{\vert\!\sim} \Theta}{A \vee B, \Gamma \mathbin{\vert\!\sim} \Theta} \, \text{L}\vee \qquad \frac{\Gamma \mathbin{\vert\!\sim} A, B, \Theta}{\Gamma \mathbin{\vert\!\sim} A \vee B, \Theta} \, \text{R}\vee$$

$$\frac{\Gamma \mathbin{\vert\!\sim} A, \Theta}{\neg A, \Gamma \mathbin{\vert\!\sim} \Theta} \, \text{L}\neg \qquad \frac{A, \Gamma \mathbin{\vert\!\sim} \Theta}{\Gamma \mathbin{\vert\!\sim} \neg A, \Theta} \, \text{R}\neg$$

Note that $\mathbin{\vert\!\sim}_0$ and $\mathbin{\vert\!\sim}$ here relate multisets. I treat things in this manner in order to avoid assuming *any* structural features absent permutation. I call $\mathbb{L}$ here NM-MS since its consequence relation is given by a <u>N</u>on-<u>M</u>onotonic <u>M</u>ulti-<u>S</u>uccedent sequent calculus.

Next I rehearse some important results for NM-MS.[4]

Theorem 1 *If $\Gamma \mathbin{\vert\!\sim} \Theta$ may be arbitrarily weakened with atoms, then it may be arbitrarily weakened with logically complex sentences:*

$$\forall \Delta_0, \Lambda_0 \subseteq \mathcal{L}_0(\Delta_0, \Gamma \mathbin{\vert\!\sim} \Theta, \Lambda_0) \Leftrightarrow \forall \Delta, \Lambda \subseteq \mathcal{L}(\Delta, \Gamma \mathbin{\vert\!\sim} \Theta, \Lambda).$$

Proof. ($\Leftarrow$) is immediate. ($\Rightarrow$) is proven by induction on the complexity of $\Delta \cup \Lambda$ where complexity is understood in terms of the complexity of the most complex sentences in $\Delta \cup \Lambda$. □

---

[3] The rules are the same as Ketonen uses. The rules with two top sequents are additive; the rules with a single top-sequent are multiplicative. These are sometimes called "mixed" or "assorted" rules/connectives (see e.g. Dicher, 2016). It is similar to the system called G3cp discussed in (Negri, Von Plato, & Ranta, 2008, ch. 3) with a more standard treatment of negation and material axioms. As is well known, these rules are equivalent to the multiplicative and additive rules of linear logic given monotonicity and contraction (Girard, 2011).

[4] Many of these results have full proofs worked out in (Girard, 2011; Negri et al., 2008).

A similar result is in the offing, namely that NM-MS preserves contraction.

Theorem 2 *If $\Gamma \mid\sim \Theta$ allows contraction of atomic sentences, then it allows contraction of logically complex sentences.*

Proof. One direction is trivial, the other direction is provided by induction on the complexity of the contracted sentence. □

Since it is well known that the rules featured above are equivalent to both the additive and multiplicative rules of linear logic given contraction and monotonicity, we can actually locate the condition needed for our logic to be supra-classical.

Definition 4 *We say that $\mid\sim_0$ obeys Containment (CO) if*

$$\forall \Delta, \Lambda \subseteq \mathcal{L}_0 (\Delta, p \mid\sim_0 p, \Lambda),$$

i.e. if we have $\forall q \in \mathcal{L}_0 (q \mid\sim_0 q)$ and all such sequents may be arbitrarily weakened; the fragment carved out by this stipulation will also obviously obey contraction). In short: let us define $\mid\sim_0^{CO}$ such that $\mid\sim_0^{CO}$ obeys reflexivity $\forall q \in \mathcal{L}_0 (q \mid\sim_0 q)$, weakening and contraction. And further stipulate that no proper subset of $\mid\sim_0^{CO}$ obeys all of these conditions. A base consequence relation $\mid\sim_0$ is said to obey CO iff it includes $\mid\sim_0^{CO}$, i.e. $\mid\sim_0^{CO} \subseteq \mid\sim_0$.

Theorem 3 *If $\mid\sim_0$ obeys CO, then $\mid\sim$ is supra-classical.*

Proof. The result is well known, and is shown by establishing equivalence with Gentzen's LK in the fragment of $\mid\sim$ generated by $\mid\sim_0^{CO}$. □

Finally, the next theorem is of particular import to the sections following this one.

Theorem 4 *All rules of NM-MS are reversible: if $\Gamma \mid\sim \Theta$ would be the result of the application of a rule to $\Gamma^* \mid\sim \Theta^*$ (and possibly $\Gamma^{**} \mid\sim \Theta^{**}$) then*

$$\Gamma \mid\sim \Theta \Leftrightarrow \Gamma^* \mid\sim \Theta^* (and \quad \Gamma^{**} \mid\sim \Theta^{**}).$$

Proof. Proof is straightforward by induction on proof height. □

From this my first gloss on logical expression follows immediately. In the next section I prove that the more precise sense (in Definition 1) also holds.

Corollary 1 *NM-MS is conservative. That is*

$$\Gamma \mid\sim_0 \Theta \Leftrightarrow \Gamma \mid\sim \Theta.$$

A Multi-Succedent Sequent Calculus for Logical Expressivists

3.1 Representation Theorem

Next I show how consequence relations may be represented in NM-MS. First I show two central results concerning conjunctive and disjunctive normal forms.[5]

Proposition 1 *Let $CNF(A)$ be the conjunctive normal form representation of A. It follows that*

$$\Gamma \mathrel{\vert\!\sim} \Theta, A \Leftrightarrow \Gamma \mathrel{\vert\!\sim} \Theta, CNF(A).$$

Proof. Proof proceeds constructively. From theorem 4, we may deconstruct A until we have a number of sequents of the form: $\Gamma \mathrel{\vert\!\sim} \Theta, \Lambda_1$; $\Gamma \mathrel{\vert\!\sim} \Theta, \Lambda_2; \ldots \Gamma \mathrel{\vert\!\sim} \Theta, \Lambda_n$ where $\Lambda_i (1 \leq i \leq n)$ contains only literals. We next construct $CNF(A)$ via repeated application of R$\vee$ and R&:

$$\Gamma \mathrel{\vert\!\sim} \Theta, (\bigvee \Lambda_1) \& (\bigvee \Lambda_2) \& \cdots \& (\bigvee \Lambda_n),$$

i.e. $\Gamma \mathrel{\vert\!\sim} \Theta, CNF(A)$. □

Proposition 2 *Let $DNF(A)$ be the disjunctive normal form representation of A. It follows that*

$$A, \Gamma \mathrel{\vert\!\sim} \Theta \Leftrightarrow DNF(A), \Gamma \mathrel{\vert\!\sim} \Theta.$$

Proof. Proof is identical to the previous proposition except the sets are on the left and we construct $DNF(A)$ via L& and L$\vee$. □

Theorem 5 (Representation Theorem 1) *Let CR be a consequence relation, i.e. $CR \subseteq \mathcal{P}(\mathcal{L})^2$. Then we may specify what must be included in $\mathrel{\vert\!\sim}_0$ such that $CR \subseteq \mathrel{\vert\!\sim}$.*

Proof. Proof proceeds constructively. For each $\Gamma \mathrel{\vert\!\sim} \Theta$ in CR let us find an equivalent $CNF(A) \mathrel{\vert\!\sim} CNF(B)$. This has the form:

$$(\& \Gamma_1) \vee \cdots \vee (\& \Gamma_a) \mathrel{\vert\!\sim} (\bigvee \Theta_1) \& \cdots \& (\bigvee \Theta_b).$$

This holds just in case (for $1 \leq i \leq a$ and $1 \leq j \leq b$) $\Gamma_i \mathrel{\vert\!\sim}_0 \Theta_j$. Thus we stipulate of the base that $\Gamma_i \mathrel{\vert\!\sim}_0 \Theta_j$ for $1 \leq i \leq a$ and $1 \leq j \leq b$. If we do this for each implication in CR then we are guaranteed that $CR \subseteq \mathrel{\vert\!\sim}$. □

---

[5]Note that the results in Propositions 1 and 2 follow closely the distribution properties Girard demonstrates for different connectives in linear logic (Girard, 1987, 2011).

Dan Kaplan

Theorem 6 (Representation Theorem 2) *Let CR be a consequence relation. If CR is closed under some modest syntactic constraints,[6] then we may specify $\mathrel{|\!\sim}_0$ such that $CR = \mathrel{|\!\sim}$.*

Proof. Proof is identical to the first Representation Theorem except that the syntactic constraints on CR have it that $\mathrel{|\!\sim} = CR$. □

These results give us a way of saying exactly how to reconstruct arbitrary consequence relations using my machinery and given some modest constraints how to reconstruct them *exactly*. It is this ability to reconstruct consequence relations *exactly* that will prove most important. For what it shows is that we are able to find exactly which pre-logical implications an arbitrary implication involving logical vocabulary *expresses*. That is, what I have shown is a method for finding exactly which implications in $\mathrel{|\!\sim}_0$ are expressed by each implication in $\mathrel{|\!\sim}$. We are thus in a position to prove the following straight away.

Theorem 7 (Expressivity) *NM-MS is expressive. That is, we have*

$$\Gamma \mathrel{|\!\sim} \Theta \Leftrightarrow (\Gamma_1 \mathrel{|\!\sim}_0 \Theta_1 \bigwedge \cdots \bigwedge \Gamma_n \mathrel{|\!\sim}_0 \Theta_n).$$

for some $\Gamma_1, \Theta_1, \ldots, \Gamma_n, \Theta_n$ and arbitrary $\mathrel{|\!\sim}_0$.

Proof. Suppose $\Gamma \mathrel{|\!\sim} \Theta$ and let it be equivalent to $DNF(A) \mathrel{|\!\sim} CNF(B)$ for some A and B. This has the form:

$$(\&\Gamma_1) \vee \cdots \vee (\&\Gamma_a) \mathrel{|\!\sim} (\vee \Theta_1) \& \cdots \& (\vee \Theta_b).$$

This holds just in case (for $1 \leq i \leq a$ and $1 \leq j \leq b$) $\Gamma_i \mathrel{|\!\sim}_0 \Theta_j$. Next, let us enumerate $\langle i, j \rangle$ as $1, \ldots, n$. Then we have that:

$$\Gamma \mathrel{|\!\sim} \Theta \Leftrightarrow (\Gamma_1 \mathrel{|\!\sim}_0 \Theta_1 \bigwedge \cdots \bigwedge \Gamma_n \mathrel{|\!\sim}_0 \Theta_n). \qquad \square$$

---

[6]In a more formal account I treat representation as of *theories*. Here I characterize it in terms of consequence relations, where we are able to precisely represent a consequence relation just in case it is closed under the rules of NM-MS. In the case where we wish to treat theories instead, then a theory T must meet the following constraints: &-composition and -decomposition ($A, B \in T$ iff $A \& B \in T$), Distribution (of $\vee$ over &) ($A \vee (B \& C)) \in T$ iff $(A \vee B) \& (A \vee C) \in T$, Conditional Equivalence ($A \to B = \sigma$ is a sub-formula of $\tau \in T$ iff $\neg A \vee B = \sigma'$ is a subformula of $\tau \in T$), both De-Morgan's Equivalences (likewise defined over sub-formulae) and involution (also defined over subformulae).

A Multi-Succedent Sequent Calculus for Logical Expressivists

4 Expressing Other Features

I have so far shown how NM-MS is expressive in the sense made precise in Definition 1. Now I show how NM-MS may express particular structural features. First I introduce a schema for introducing a piece of logical vocabulary '$\boxed{\mathfrak{S}}$'.

First, let us enrich our sequent calculus by introducing a second turnstile $\vdash^{\mathfrak{S}}$. Now let $\vdash_0^{\mathfrak{S}}$ pick out some subset of $\vdash_0$. Later I will discuss principles for determining which subset, but for now I leave the details vague. We may introduce the following rules to our sequent calculus:[7]

Axiom 2: If $\Gamma \vdash_0^{\mathfrak{S}} \Theta$ then $\Gamma \vdash^{\mathfrak{S}} \Theta$.

$$\frac{A, \Gamma \vdash^{\mathfrak{S}} \Theta}{\boxed{\mathfrak{S}}A, \Gamma \vdash^{[\mathfrak{S}]} \Theta} \, L\boxed{\mathfrak{S}} \qquad \frac{\Gamma \vdash^{\mathfrak{S}} \Theta, A}{\Gamma \vdash^{[\mathfrak{S}]} \Theta, \boxed{\mathfrak{S}}A} \, R\boxed{\mathfrak{S}}$$

Lemma 1 $L\boxed{\mathfrak{S}}$ *and* $R\boxed{\mathfrak{S}}$ *are reversible rules.*

We thus have the following result.

Theorem 8 *Let $\mathfrak{Sf}$ be a structural rule. Suppose that $\mathfrak{Sf}$ is preserved (in the sense of Definition 2) and suppose further that $\vdash^{\mathfrak{S}}$ marks that structural feature exactly. We thus have: $\mathfrak{Sf}(\Gamma \vdash \Theta)$ iff $\Gamma \vdash^{\mathfrak{S}} \Theta$. It follows that $\boxed{\mathfrak{S}}$ expresses (in the sense of Definition 3) $\mathfrak{Sf}$. Thus:*

$$\boxed{\mathfrak{S}}A, \Gamma \vdash \Theta \Leftrightarrow \mathfrak{Sf}(A, \Gamma \vdash \Theta)$$
$$\Gamma \vdash \Theta, \boxed{\mathfrak{S}}A \Leftrightarrow \mathfrak{Sf}(\Gamma \vdash \Theta, A)$$

Proof. I prove only the latter biconditional since the proof of the former is identical. By supposition $\mathfrak{Sf}(\Gamma \vdash \Theta, A)$ iff $\Gamma \vdash^{\mathfrak{S}} \Theta, A$. Since it follows that our $R\boxed{\mathfrak{S}}$ rule is reversible, we have that $\Gamma \vdash^{\mathfrak{S}} \Theta, A$ iff $\Gamma \vdash \Theta, \boxed{\mathfrak{S}}A$. Thus

$$\Gamma \vdash \Theta, \boxed{\mathfrak{S}}A \Leftrightarrow \mathfrak{Sf}(\Gamma \vdash \Theta, A). \qquad \square$$

The result of the above proof is a general method for introducing logical vocabulary that is *expressive* of structural features. If the rules for the logical vocabulary's introduction are reversible and the structural feature in question is *preserved* by $\mathbb{L}$, then the logical vocabulary will *express* that structural feature. I next rehearse two specific cases of this: an operator that marks monotonicity and an operator that marks classical validity.

---

[7] Note that the rest of our sequent calculus is altered such that our other rules preserve $\vdash^{\mathfrak{S}}$. E.g. R& requires that both top sequents have either $\vdash$ or $\vdash^{\mathfrak{S}}$ (I do not allow mixed cases).

4.1 Capturing Monotonicity '$\boxed{M}$' and Classicality '$\boxed{K}$'

The rules for monotonicity have the following form:

Axiom 2: If $\forall \Delta, \Lambda \subseteq \mathcal{L}_0(\Delta, \Gamma \mathrel{\vert\!\sim}_0 \Theta, \Lambda)$ then $\Gamma \mathrel{\vert\!\sim}^M \Theta$.

$$\frac{A, \Gamma \mathrel{\vert\!\sim}^M \Theta}{\boxed{M}A, \Gamma \mathrel{\vert\!\sim}^{[M]} \Theta} \; \text{L}\boxed{M} \qquad \frac{\Gamma \mathrel{\vert\!\sim}^M \Theta, A}{\Gamma \mathrel{\vert\!\sim}^{[M]} \Theta, \boxed{M}A} \; \text{R}\boxed{M}$$

I have already show in Theorem 1 that weakening is preserved by the rules of NM-MS. It therefore follows that:

Corollary 2 $\boxed{M}$ *expresses weakening/monotonicity. That is,*

$$\boxed{M}A, \Gamma \mathrel{\vert\!\sim} \Theta \Leftrightarrow \forall \Delta, \Lambda (\Delta, A, \Gamma \mathrel{\vert\!\sim} \Theta, \Lambda)$$
$$\Gamma \mathrel{\vert\!\sim} \Theta, \boxed{M}A \Leftrightarrow \forall \Delta, \Lambda (\Delta, \Gamma \mathrel{\vert\!\sim} \Theta, A, \Lambda)$$

This means that we may expand NM-MS (our $\mathbb{L}$) in order to mark *in the object language* which implications are persistent under arbitrary weakenings. Next, I demonstrate the same for "classicality", i.e. develop an operator that marks implications that are valid classically.

Axiom 2: If $\Gamma, p \mathrel{\vert\!\sim}_0 p, \Theta$ then $\Gamma, p \mathrel{\vert\!\sim}^K p, \Theta$ (where Γ, Θ may be possibly empty).

$$\frac{A, \Gamma \mathrel{\vert\!\sim}^K \Theta}{\boxed{K}A, \Gamma \mathrel{\vert\!\sim}^{[K]} \Theta} \; \text{L}\boxed{K} \qquad \frac{\Gamma \mathrel{\vert\!\sim}^K \Theta, A}{\Gamma \mathrel{\vert\!\sim}^{[K]} \Theta, \boxed{K}A} \; \text{R}\boxed{K}$$

Again, I have already shown in Theorem 3 that classicality is a feature NM-MS preserves. Thus any sequent which is derived from atomic sequents which are part of the CO (cf. Definition 4) fragment of $\mathrel{\vert\!\sim}_0$ (regardless of whether $\mathrel{\vert\!\sim}_0$ actually obeys CO) will be classically valid.

Corollary 3 *Let $\vdash_{LK}$ be the consequence relation instantiated by Gentzen's LK minus the rules for quantifiers (and with $\wedge$ substituted with &, etc.). Then $\boxed{K}$ expresses classical validity, that is:*

$$\boxed{K}A, \Gamma \mathrel{\vert\!\sim} \Theta \Leftrightarrow A, \Gamma \vdash_{LK} \Theta$$
$$\Gamma \mathrel{\vert\!\sim} \Theta, \boxed{K}A \Leftrightarrow \Gamma \vdash_{LK} \Theta, A$$

A Multi-Succedent Sequent Calculus for Logical Expressivists

There are of course many further possibilities for such '⊖' operators. We may also introduce vocabulary for expressing inference that obey contraction, transitivity + weakening, more restricted weakening principles, and perhaps more.[8]

5 Some Defective Cases

So far I have introduced a more precise criterion for understanding logical expressivism and in particular for understanding how *structural* features of inference might be expressed. I then introduced a system that was not only *expressive* in this sense, but also successfully preserved and expressed several important structural features. In order to appreciate exactly what I am up to, however, it will be useful to look at some cases where each of these criteria fail.

Example 1 The multiplicative rules of linear logic are *not expressive*. I show that this is the case for the multiplicative conjunction $\otimes$:

$$\frac{\Gamma, A, B \vdash \Theta}{\Gamma, A \otimes B \vdash \Theta} \, L\otimes \qquad \frac{\Gamma \vdash \Theta, A \quad \Delta \vdash \Lambda, B}{\Gamma, \Delta \vdash \Theta, \Lambda, A \otimes B} \, R\otimes$$

It is sufficient to show a case where the logic does not express particular implications in $\vdash_0$. Notice that there are potentially two ways to derive $p \otimes q \vdash p \otimes q$ where $p, q \in \mathcal{L}_0$:

$$\frac{\dfrac{p \vdash p \quad q \vdash q}{p, q \vdash p \otimes q} \, R\otimes}{p \otimes q \vdash p \otimes q} \, L\otimes \qquad \frac{\dfrac{p, q \vdash q}{p \otimes q \vdash q} \, L\otimes \quad \vdash p}{p \otimes q \vdash p \otimes q} \, R\otimes$$

Since the atomic sequents used to start each proof tree are different (in fact they are entirely different), it's possible that $\vdash_0$ includes one and $\vdash'_0$ includes the other and thus the presence of $p \otimes q \vdash p \otimes q$ does not guarantee the presence of either. In this sense, logics which include '$\otimes$' are not expressive in the relevant sense.

---

[8]Makinson for example considers a consequence relation which is supra-classical, monotonic, and obeys transitivity (Makinson, 2003, 2005). We could introduce an operator to express exactly this consequence relation along with some other consequence relations discussed therein.

It is also possible to find counter-examples to Sf-Preservation and Sf-Expression. Even using the rules of NM-MS such counter-examples will arise:

Example 2 Suppose we want to introduce an operator '$\boxed{R}$' to mark instances of reflexivity, i.e. $\phi \mathrel{\vert\!\sim} \phi$. Then the rules for introducing such an operator should probably have the form:

Axiom 2: If $p \mathrel{\vert\!\sim}_0 p$ then $p \mathrel{\vert\!\sim}^R p$.

$$\frac{A, \Gamma \mathrel{\vert\!\sim}^R \Theta}{\boxed{R}A, \Gamma \mathrel{\vert\!\sim}^{[R]} \Theta} \text{ L}\boxed{R} \qquad \frac{\Gamma \mathrel{\vert\!\sim}^R \Theta, A}{\Gamma \mathrel{\vert\!\sim}^{[R]} \Theta, \boxed{R}A} \text{ R}\boxed{R}$$

Unfortunately, it is easy to show that NM-MS fails to preserve reflexivity and thus fails to express it. For example $A\&B \mathrel{\vert\!\sim} A\&B$ is clearly an instance of reflexivity and thus we should want $A\&B \mathrel{\vert\!\sim} \boxed{R}(A\&B)$. But clearly $A\&B \mathrel{\vert\!\sim} A\&B$ must be derived from $A, B \mathrel{\vert\!\sim} A$ and $A, B \mathrel{\vert\!\sim} B$, neither of which are instances of reflexivity.[9]

There will therefore be logics which in general fail to be expressive and even among those that are expressive there will be structural features that fail to be preserved and thus expressed. Deciding how expressive one wants one's logic to be and which structural features ought to be preserved are therefore *not* independent questions.

6 Conclusion

In this paper I introduced a precisification of the notion of "logical expression". I also argued for two additional criteria for understanding when a structural feature is preserved and expressed. With these criteria in hand I introduced a system NM-MS. NM-MS is a sequent calculus without any structural features or restrictions placed on it. I showed that NM-MS is *expressive* in the precise, technical sense I argued for and I also exhibited some other interesting features it possesses. I next introduced some machinery for

---

[9]Though they are both found in the region of the consequence relation which allows reflexivity *together with weakening*, hence why we are able to have an operator to mark classicality.

It is also worth remarking that the above might also fail for independent reasons. For example, if we are able to derive $A\&B \mathrel{\vert\!\sim} A\&B$, then we could also derive $A\&B \mathrel{\vert\!\sim} B\&A$, but is the latter here an instance of the structural feature of reflexivity? It is not obvious that we should think so. In general, even when a sequent calculus preserves reflexivity, it needn't generate *only* reflexive sequents from the reflexive fragment of its axioms.

A Multi-Succedent Sequent Calculus for Logical Expressivists

marking and thus expressing structural features in NM-MS where they occur. I showed that monotonicity and classicality are two features that may each be preserved and expressed in this way. I closed by exhibiting some cases where a logic *fails* to be expressive or *fails* to preserve and/or express a structural feature. The goal of the paper was to make the thesis of logical expressivism more precise and to introduce a logic which is actually expressive in the relevant sense. My hope is that providing such an account might help illuminate exchanges within the philosophy of logic: to proponents of expressivism, a clearer doctrine and a logic to call their own, and to those opposed, a clearer target.

References

Brandom, R. (1994). *Making It Explicit: Reasoning, Representing, and Discursive Commitment*. Cambridge, MA: Harvard University Press.

Brandom, R. (2008). *Between Saying and Doing: Towards an Analytic Pragmatism*. Oxford University Press.

Dicher, B. (2016). A proof-theoretic defence of meaning-invariant logical pluralism. *Mind*, *125*(499), 727–757.

Girard, J.-Y. (1987). Linear logic. *Theoretical computer science*, *50*(1), 1–101.

Girard, J.-Y. (2011). *The Blind Spot: Lectures on Logic*. Zürich: European Mathematical Society.

Hlobil, U. (2016). A nonmonotonic sequent calculus for inferentialist expressivists. In P. Arazim & M. Dančák (Eds.), *The Logica Yearbook 2015*. College Publications.

Makinson, D. (2003). Bridges between classical and nonmonotonic logic. *Logic Journal of IGPL*, *11*(1), 69–96.

Makinson, D. (2005). *Bridges From Classical to Nonmonotonic Logic*. London: King's College.

Negri, S., Von Plato, J., & Ranta, A. (2008). *Structural Proof Theory*. Cambridge University Press.

Peregrin, J. (2014). *Inferentialism: Why Rules Matter*. London: Palgrave Macmillan.

Dan Kaplan
University of Pittsburgh
USA
E-mail: dan.kaplan@pitt.edu

Logical Dialogues in Abstract Argumentation Frameworks

HANNA KARPENKO[1] AND OLIVIER ROY

Abstract: We show that Lorenzen & Lorenz's dialogical logics Lorenzen and Lorenz (1978) with the mechanism of repetition ranks can be represented in Dung abstract argumentation framework. More specifically, we show that Proponent's winning strategies for a particular formula correspond to that formula being part of the preferred extension in a corresponding arguments frame. We focus on the dialogical approach to First-Order Logic, as developed in Clerbout (2014a).

Keywords: Dialogical logics, Abstract argumentation, Winning strategies, Game players, Extensive form of a dialogue

1 Introduction

Dialogical Logic is built on the idea that interaction in the form of a dialogue lies at the foundations of laws of reasoning. Theories of Abstract Argumentation depart from the idea that interaction between two arguing sides is the base of various instances of non-monotonic reasoning. Resulting formalisations differ a lot between themselves. One introduces explicitly two players, whereas the other does not. The second does not describe the behaviour of logical operators, whereas local rules of the first are dedicated to them. Yet, both seem to assume that at least two arguing sides need to exist for use of reasoning to take place. This assumption leads one to suspect these two systems to be alike. We try to grasp formally the alikeness of these two frames.

---

[1] I wish to thank Shahid Rahman, Andriy Vasylchenko, Volodymyr Navrotskyi and Logica17 participants.

2 Dialogical logics and abstract argumentation frameworks

2.1 Dialogical logics

The original motivation behind this frame is to give logical constants interpretations which show their place in an *argumentative debate*. Formally, this has been done by constructing dialogical *games* for logical validity.

We focus on the dialogical approach to First-Order Logic, as developed in (Clerbout, 2014a). Dialogical games are played by two players, Opponent (**O**) and Proponent (**P**). **P** defends a thesis stated at the beginning of the debate. The different moves of the games fall into two categories: *posits and questions, attacks and defenses*. A posit is an affirmative statement. A question is an attack of a player on a posit made by their adversary, which asks for specifications. A way to make a question is to ask to state a part of complex formula or to replace a quantified formula by its instantiation. An attack is a challenge of a complex formula. For instance, an attack on a statement which consists of a formula whose main connective is $\wedge$ is to ask to state its first or second conjunct. A defense is a statement of a part of a complex formula or statement of instantiation of a quantified formula, which is carried out as response to a certain attack.

Syntactically, dialogical logic extends the usual FOL vocabulary with names of two players (**O**) and (**P**), as well as signs for questions (?) and posits (!). The role of **P** is to state initial thesis and try to defend it. The role of **O** is to prevent **P** from defending the thesis. We use pronouns "he" and "she" for **P** and **O** respectively. In case where it can be any of players, we use "she". We use labels **X** and **Y** when the identity of players does not matter.

The meaning of logical expressions is spelled out in terms of specific attack and defense rules for connectives and operators.

The rules are divided into structural and local rules. Local rules for propositional connectives are:

Posit	$X!\ A \wedge B$		Posit	$X!\ A \vee B$	
Attack	$Y?\ \wedge_1$	$Y?\ \wedge_2$	Attack	$Y?\ \vee$	
Defence	$X!\ A$	$X!\ B$	Defence	$X!\ A$	$X!\ B$

Posit	$X!\ A \rightarrow B$	Posit	$X!\ \neg A$
Attack	$Y!A$	Attack	$Y!A$
Defence	$X!\ B$	Defence	-

Representing Dialogues as Argumentation Frameworks

An *X-terminal play* is a play where the last move is an X-move, and where there are no moves which Y can make according to dialogue rules.

A *repetition rank* for a player X is a number of times this player could react on the same move of her adversary. It is chosen from positive integers by each player.

And the four structural rules for classical logic are:

1. **Starting Rule**: The dialogue begins by **P**'s stating the thesis at move 0. Then the choices of repetition ranks by **O** and **P** follow at moves 1 and 2.

2. **Development Rule**: After the repetition ranks have been chosen, any move in a play is either challenge or defense, played in relation to a move that precedes them. A player can defend herself against any attack of her adversary, or make an attack herself. (For intuitionistic dialogues the rule is modified so **P** can defend himself only against last undefended attack by **O**. This is the only difference with classical structural rules).

3. **Formal Rule**: **P** has a right to state an atomic sentence only if it was stated beforehand by **O**.

4. **Winning Rule**: A player X wins a play if the play is X-terminal.

A *dialogue*, or *dialogical game* over ϕ, noted $D(\phi)$, is a set of all plays according to local and structural rules for a given logic, such that ϕ is a thesis of these plays. The *extensive form* $\mathfrak{E}(\phi)$ of $D(\phi)$ is the tree representation of $D(\phi)$. Nodes are labeled with moves, the root is labeled with the thesis, paths in $\mathfrak{E}(\phi)$ are linear representations of plays and paths that can not be extended represent terminal plays in $D(\phi)$.[2]

---

[2] The definition of extensive form is taken from (Clerbout, 2014a).

Hanna Karpenko and Olivier Roy

	O			**P**	
				$!\,((p \wedge \neg p) \to q)$	0
1	$!n := 1$			$!m := 2$	2
3	$!p \wedge \neg p$	0			
5	$!\neg p$		3	$?\neg p$	4
7	$!p$		3	$?p$	6
			5	$!p$	8

Table 1: A table showing a **P**-winning play in the game over $(p \wedge \neg p) \to q$. Two exterior columns show the number of the current move. Two interior columns show the move which is under attack by the current move. Two columns in between show the action by a player and the content of that action at the current move. At move 0 **P** affirms a thesis. At moves 1 and 2 players choose their repetition ranks. At move 3 **O** attacks the thesis by affirming its antecendent. The further development of the play can be understood relying on structural and local rules.

In figure 1, all nodes of depth 1 and 2 show possible choices of a repetition rank. All the nodes of higher depth show possible moves according to local and structural rules. Notice that the branches that follow the node **P**!m:=1 are **O**-terminal : as P could not attack the conjunction the second time, he could not win at all if he chooses his repetition rank m:=1.

A *strategy* for a player **X** in a dialogical game over $\phi, D(\phi)$, is a function which to each **Y**-move in a non-terminal play assigns an **X**-move. An **X**-strategy is *winning* if **X**'s win is independent from ways for **Y** to play. For example, a play described in 1 can be seen as describing **P**'s strategy for a game over $(p \wedge \neg p) \to q$: the only way for **O** to play differently is to choose a different repetition rank, and to counterattack the thesis again and again. But one can see that in that case **P** can just copy-paste his responses from the given play. The way **O** plays does not thus influence the outcome of the play.

2.2 Dung's abstract argumentation frameworks

We use the following definitions of Argumentation Framework system, proposed by P.M.Dung in Dung (1970):

An *argumentation framework* is a pair: $AF = \langle AR, attacks \rangle$, where AR is a set of arguments, and $attacks$ is a binary relation on AR, i.e. $attacks \subseteq AR \times AR$.

Representing Dialogues as Argumentation Frameworks

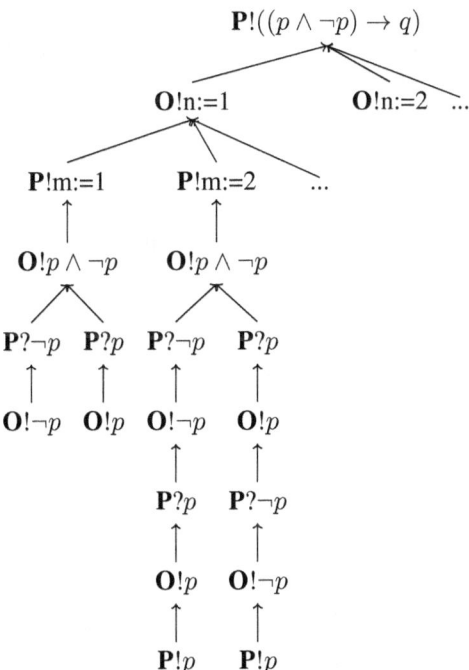

Figure 1: The extensive form of $D((p \land \neg p) \to q)$

1. An argument $A \in AR$ is *acceptable* with respect to set S of arguments iff for each argument $B \in AR$: if B attacks A, then B is attacked by S.

2. A conflict-free set of arguments S is *admissible* iff each argument in S is acceptable with respect to S.

A *preferred extension* on an argumentation framework AF is a maximal (with respect to set inclusion) admissible set of AF. As will become clear later, Argumentation Frameworks we are interested in will always produce exactly one preferred extension. We thus use functional notation $Pref(AF())$ to denote the preferred extension of an Argumentation Framework. An argumentation framework is *well-founded* iff there exists no infinite sequence $A_0, A_1, ..., A_n$, such that for each i, A_{i+1} attacks A_i.

As an example of an Argumentation Framework we can think of the discussion of two people consisting of the following arguments:

A You made an intervention to another country without any serious reason.

B I did it to protect rights of people living in it.

C There was no violation of their rights which preceded your intervention.

In this case, $AR = \{A, B, C\}$, $attacks = \{(A, B), (B, A), (C, B)\}$ and $Pref(AF) = \{A, C\}$.

3 Two possible ways to represent an Argumentation Framework for a dialogue

The main result of this paper is:

> **P** has a winning strategy in a dialogue over a thesis ϕ iff the thesis labeled by **P** is found in the preferred extension of the argument over ϕ in Dung's framework.

We use the basic intuition according to which an extensive form of a dialogue or its fragment can be seen as AFs. The elements of a set of arguments in a corresponding Argumentation Framework will always be players' posits or questions labeled by their names, such as **P**!$p \to p$, **O**?$\wedge_1$.

Representing Dialogues as Argumentation Frameworks

We will use $AF(D(\phi))$ notation to say that an Argumentation Framework is constructed for the dialogue over ϕ. One can think of constructing AF out of plays using extensive forms in two ways:

1. As a fragment of $\mathfrak{E}(\phi)$ which describes **P**-winning strategy in a game over ϕ.

2. As the whole $\mathfrak{E}(\phi)$

Let's firstly look at 2.

3.1 AF as whole extensive form

It turns out that the result depends on whether in the set *attacks* we include only *direct attacks* or *indirect attacks* as well. Direct attack is an attack which follow the argument immediately: in the example from the previous section, (B, A), and (C, B) are both direct attacks. Indirect attack is an attack on arguments that defend the argument against direct attacks or indirect attacks which came earlier after direct defense. If we had an argument D which attacked C, this same argument would be an indirect attack on A.

To see why the result would not hold should we include *indirect attacks*, take the formula $((p \land \neg p) \to q)$. One of the branches of its extensive form will result from both players' choices of their repetition ranks to be 1. Take one of two leftmost branches of figure 1 to see how the branch in question will look like in $\mathfrak{E}((p \land \neg p) \to q)$. There is an **O**-labeled leaf which represents an indirect attack on the root that is not defended by **P**. The root of the AF cannot be admissible, and, therefore $\notin Pref(AF(D(\phi)))$. On the other hand, there is clearly a winning strategy for **P** in a dialogue over $(p \land \neg p) \to q$ in both classical and intuitionistic logic.

However, if we include only *direct attacks* under *attacks*, then the same counter-example does not hold: take again figure 1: even if **P**!m:=1 does not count as admissible argument, there are other **P**-labeled moves that answer **O**!n:=1 and that count as admissible arguments. Therefore, **P**!$(p \land \neg p) \to q$ will still be in $Pref(AF(D((p \land \neg p) \to q)))$.

Yet the whole extensive form contains all the choices of repetition ranks by **O** and **P**, which increase infinitely. This makes the evaluation of our main argument unnecessarily complicated. Another possibility is to consider only fragments of extensive form which represent **P**-winning strategies.

3.2 AF as a fragment of $\mathfrak{E}(\phi)$ describing a P-strategy

Let us use following definitions of an extensive form of a strategy and the subsequent fact given in Clerbout (2014a).

Let s_x be a strategy of player **X** in $D(\phi)$ of extensive form $\mathfrak{E}(\phi)$. The extensive form of s_x is the fragment $\mathfrak{S}_x$ of $\mathfrak{E}(\phi)$, such that the root of $\mathfrak{E}(\phi)$ is the root of $\mathfrak{S}_x$, and any immediate successor of node t which is associated with an **X**-move in $\mathfrak{E}(\phi)$, is an immediate successor of t in $\mathfrak{S}_x$.

Fact 1 *For any node t which is associated with a **Y**-move in $\mathfrak{E}(\phi)$, if t has at least an immediate successor in $\mathfrak{E}(\phi)$, then t has exactly one immediate successor in $\mathfrak{S}_x$, namely the one labeled with the **X**-move prescribed by s_x.*

For each fragment $\mathfrak{S}_x$ of $\mathfrak{E}(\phi)$, we take it as an AF over ϕ where AR is set of all moves by **P** and **O** in $\mathfrak{S}_x$, and $attacks$ is a set of couples, such that an arbitrary couple is constituted of nodes t_k and t_{k+1}, where t_k and t_{k+1} are moves by the players, and t_{k+1} is an immediate successor of t_k in the $\mathfrak{E}(\phi)$.

The two following theorems are used subsequently in the development of our argument:

Theorem 1 *Let $\mathfrak{S}_x$ be an extensive form of **P**-strategy S_p. If S_p is winning, then every leaf of $\mathfrak{S}_x$ is associated with affirmation of atomic formula by **P**.*[3]

Theorem 2 *Determinacy There is a **P**-strategy in $D(\phi)$ iff there is no **O**-strategy in $D(\phi)$.*[4]

Now we can give the exact formulation of our main result:

Theorem 3 *P has a winning strategy in a dialogical game over a thesis ϕ iff there exists $AF(D(\phi), Sp)$ such that $P!\phi \in Pref(AF(D(\phi), S_p))$, where $AF(D(\phi), S_p)$ is an Argumentation Framework constructed from S_p fragment of $D(\phi)$.*

4 Propositional logic

Let us consider the case of propositional (classical or intuitionistic) logic.

For convenience, we will use notions **O**-labeled and **P**-labeled node for nodes which respectively represent **O** and **P**-moves.

---

[3] Find proof in Clerbout (2014b).
[4] Proof is in Clerbout (2014b).

Any fragment of $\mathfrak{E}(\phi)$ which keeps all **O**-labeled nodes and exactly one **P**-labeled following node for each **O**-labeled non-terminal node, is considered a **P**'s strategy.

$D^{n,m}(\phi)$ stands for *fragment of $D(\phi)$ reduced to all plays where **O** chooses her rank to be n, and **P** chooses his rank to be m*. D^n stands for *fragment of $D(\phi)$ reduced to all plays where **O** chooses her rank to be n*

Proof. For the case of Propositional Logic we can use the following result:

Theorem 4 *P has a winning strategy in $D(\phi)$ iff he has a winning strategy in $D^{1,2}(\phi)$. (Clerbout, 2014a)*

We thus consider only strategies constructed out of $D^{1,2}(\phi)$ fragments of the dialogue. Let us prove the wanted result by induction on the number of attacks (the length of the tree). Take an arbitrary **P**-strategy, S_{n_p} with its extensive form $\mathfrak{S}_{n_p}$ in a game over ϕ.

1. Let us establish the max. number of attacks = 3.[5]

    By the construction of dialogue, the first attack is **O**-labeled:

    where X is either ? or ! and y is the corresponding specification (either part of conjunction which is asked, or a question for disjunction, or else an antecedent stated). Thus, neither the strategy which can be described by this tree will be a **P**-winning strategy, nor all leaves of such tree will be **P**-labeled, which means that there are undefended **O**-attacks.

2. Let us assume that the result holds for a number of attacks $\leq n$. It could happen in two cases:

    (a) All the leaves of the $\mathfrak{S}n_p$ are **P**-labeled.

    (b) There is at least one **O**-labeled leaf in $\mathfrak{S}n_p$.

    Let us suppose the case is (a). We add to the longest branch a node, which is, by construction, **O**-labeled node. There is one undefended

---

[5]First two attacks will constitute choices of repetition ranks.

O-labeled attack. Automatically, $\mathbf{P}!\phi \notin Pref(AF(D(\phi)), Sn_p))$. By Theorem 1, the corresponding described **P**-strategy is not **P**-winning. Suppose the case is (b). This case contains the following subcases:

(i) There is only one **O**-labeled leaf, which is found in the branch of maximal length in the fragment $\mathfrak{S}n_p$. Then, when we add one more node, by construction, it would be **P**-labeled. Now, all nodes are **P**-terminal. That means, that

$$\mathbf{P}!\phi \in Pref(AF(D(\phi)), Sn_p)).$$

In the fragment, each **O**-labeled move is associated with **P**-labeled move and all plays represented are **P**-winning. As by construction, all **O**-labeled choices are in the tree, **P** has a winning strategy.

(ii) There is more than one **O**-labeled leaf. Then, no matter how the leaf of the maximal branch is labeled, the reasoning can be reduced to these of assumptions of inductive step.

(iii) There is only one **O**-labeled leaf, which is not the leaf of maximal branch. Then, the reasoning can be reduced to that of assumptions of inductive step. □

5 FOL

5.1 Semi-decidability of FOL dialogues

For the case of propositional classical and intuitionistic logics, we used the result according to which **P** has a winning strategy in $D(\phi)$ iff **P** has a winning strategy in $D^{1,2}(\phi)$. The following considerations allowed us to restrict our attention to the $D^{1,2}$ dialogues.

Clerbout (2014b) proves two results :

1. There is a winning **P**-strategy in $D(\phi)$ iff there is one in $D^1\phi$.

2. There is no positive integer n such that, for any ϕ, there is a winning **P**-strategy in $D(\phi)$ iff there is one in $D^{1,n}(\phi)$.

We are thus allowed to restrict ourselves to fragment D^1, but we are not allowed to restrict ourselves only to strategies resulting from one certain rank chosen by **P**.

Representing Dialogues as Argumentation Frameworks

At the same time, FOL and, correspondingly, FOL dialogues are semi-decidable. For dialogical logic that means that, if there is a **P**-winning strategy, then it can be discovered after trying a finite number of choices of repetition ranks.

5.2 Other infinitary rules

The choices of repetition ranks are not the only infinitary rules in FOL games. Also, the rules for quantifiers produce an infinite number of branches in extensive forms of games, and, consequently, of players' strategies.

Rules for quantifiers could be formulated as follows :

Posit	**X**! $\forall x A(x)$	Posit	**X**! $\exists x A(x)$
Attack	**Y**? x/a	Attack	**Y**? x/
Defence	**X**! A(a)	Defence	**X**! A(a)

So, in the case of the universal quantifier, it is an attacking player who chooses the individual constant to substitute the bounded variable. In the case of existential quantifier, it is the defending player.

It is clear that sometimes these new infinitary rules provoke infinite number of branches within a single **P**-strategy. It is equally clear that the new branches constituting the infinite number will be only alphabetical variants of already existing branches. Now at least the proof of the following part of the result can be adapted from propositional case:

Theorem 5 *If **P** has a winning strategy in a dialogical game over a FOL thesis ϕ, then there exists $AF(D(\phi), Sp)$ such that **P**! $\phi \in Pref$ of $AF(D(\phi), S_p)$.*

Proof. After finding a repetition rank m at which **P** could win, keeping the fragment $D^{1,m}(\phi)$, and keeping only one branch from those which are alphabetical variants of one another, we could directly apply the same reasoning as in the propositional case. □

As for the inverse direction, namely:

Theorem 6 *If there exists $AF(D(\phi), Sp)$ such that ϕ is a FO-formula and **P**!$\phi \in Pref$ of $AF(D(\phi), S_p)$, then **P** has a winning strategy in a dialogical game over ϕ.*

...the following argument holds:

Proof. We can always consider $D^1(\phi)$ instead of $D(\phi)$. Moreover, by the rules of construction of **P**-strategies and AF-frameworks over them, there would be no infinite number of **P**-nodes which are successors of the same **O**-node. Length of any given branch of the strategy will not be infinite either. Thus only infinite branches will be ones generated by **O**'s defense on **P**'s attack on **O**!$\exists x \psi(x)$, or **O**'s attack on **P**!$\forall x \psi(x)$, where $\psi(x)$ is a sub-formula of ϕ. But if **P**!$\phi \in Pref(AF(D(\phi), Sn_p))$, then **P**!$\phi$ is admissible. This means that all **O**-attacks on ϕ are **P**-defended. And that means that all branches of the tree are **P**-terminal. All plays are thus **P**-winning. As all **O**-moves from the $D(\phi)$ are included, it is not possible for **O** to play otherwise in $D(\phi)$. The considered strategy is this **P**-winning. □

6 Well-foundedness and finiteness of argumentation frameworks

Recall two definitions from (Dung, 1994): An *AF* is **well-founded** iff there exists no infinite sequence $A_0, A_1, ..., A_n$ such that for each i, A_{i+1} attacks A_i. An *AF* is **finitary** iff for each argument A, there are only finitely many arguments in AR which attack A.

We have thus two following result concerning FOL - dialogues :

1. Any $\mathfrak{S}n_x$ is an example of *well-founded AF*. As explained before, repetition ranks always guarantee finiteness of a play. And branches in extensive forms in strategies are always represented by plays.

2. Some $\mathfrak{S}n_x$ using the vocabulary of FOL are examples of non-finitary AF. Such examples are strategies where the infinite number of branches if produces by **O** moves which are alphabetical variants of each other. Or, by **O**'s choices of repetition ranks, if we consider $D(\phi)$ instead of $D^1(\phi)$.

From Dung (1970) we know that if an AF is well-founded, then its grounded, preferred and stable extensions coincide, which can be applied to extend the discussed result.

7 Conclusion

The present work cannot be spoken of as the first one comparing dialectical frameworks with abstract argumentation. The closest version, maybe, is the number of proofs given in Prakken (2005). We placed ourselves in more strict dialogical settings to be opposed to larger dialectical settings of Prakken. Equally, the mechanism of repetition ranks and the application of the result to FOL are topics which are absent in the mentioned article.

The result points on the fact that there is the same argumentative structure for both frames. It also shows that non-monotonic and monotonic logics mostly share the same argumentative structure.

The consideration of extensive forms of dialogues also points at one thing. Recall the fact that one can not prove the result considering the whole extensive forms and indirect attacks. This seems to point at the fact that playing/arguing according to the rules does not imply playing/arguing strategically. It seems thus that Argumentation Frameworks are mostly interested in the strategical level of reasoning. In Dialogues, this level is displayed by players' strategies.

References

Clerbout, N. (2014a). First-order dialogical games and tableaux. *Journal of Philosophical Logic*, *43*, 785–801.

Clerbout, N. (2014b). *La Sémantique Dialogique* (Unpublished doctoral dissertation). L'Université de Lille, Lille.

Dung, P. M. (1970). On the acceptability of arguments and its fundamental role in nonmonotonic reasoning, logic programming and n-person games. *Artificial Intelligence*, *77*, 321–357.

Lorenzen, P., & Lorenz, K. (1978). *Dialogische Logik*. Darmstadt: Wissenschaftliche Buchgesellschaft.

Prakken, H. (2005). Coherence and flexibility in dialogue games for argumentation. *Journal of Logic and Computation*, *15*, 1009–1040.

Hanna Karpenko
E-mail: `hannakarpe@gmail.com`

Olivier Roy
Bayreuth University
Germany
E-mail: `olivier.roy@uni-bayreuth.de`

"I Asked You to Mail that Letter, Not to Burn It," An Illocutionary Logical Analysis of Directive Acts and Arguments

JOHN T. KEARNS

Abstract: *Illocutionary logic* is the logic of speech acts, or language acts. The "units" of significant speech are *illocutionary acts*, which are typically constituted by someone's using a sentence or sentential clause to perform a certain job, like making an assertion or denial, giving an order, or making a promise. To use that sentence or clause to say something meaningful is to perform a *locutionary act*, and to perform the locutionary act in a certain manner, or with a certain force which determines what job is carried out is to perform an illocutionary act. The most familiar logical theories focus on *statements*, which are *assertive locutionary acts*, though there are also theories of the *assertive illocutionary arguments* which are composed from assertions, denials, and suppositions. A *directive illocutionary act* orders, or requests, or recommends that its addressee perform or not perform a certain action. A *directive locutionary act* is a *plan*, like "Mark, close the door," which is performed with a suitable force to constitute the illocutionary act. Just as there are logical theories of assertive locutionary acts and assertive illocutionary acts and arguments, so there are (or can be) logical theories of directive locutionary acts and directive illocutionary acts and arguments. This paper explores some features of directive acts and arguments.

Keywords: Speech act, Language act, Speech act logic, Illocutionary acts, Illocutionary arguments, Directive acts, Illocutionary logic

1 Illocutionary acts

Illocutionary logic is a study of speech acts, or language acts, and of arguments composed from language acts. A speech act, or language act, is a meaningful act that someone performs by saying something, or writing something, or thinking something using words and sentences. *Illocutionary acts* are one kind of language act, such an act is typically one in which someone uses a sentence or sentential clause to perform a certain job, like making a promise or making a request, threatening someone or giving her advice, asserting that something is the case, or asking someone what time it is.

Illocutionary acts are the "units" of speech or language, they are the complete, concrete language acts that people perform when they address or write to one another, or when they are thinking things through for themselves. Assertions, denials, and acts of supposing a statement to be the case, or not to be the case, are illocutionary acts; they are examples of *assertive illocutionary acts*. A straightforward assertion might be constituted by someone's using a sentence to represent things being a certain way, accepting that this is how things are. A person can make an assertion when she is alone, or she can address her assertion to someone else, endorsing the statement that she asserts.

Assertive illocutionary acts are one of the five categories of illocutionary acts in John Searle's taxonomy (this is found in Searle, 1969, 1985). *Directive (illocutionary) acts* constitute another category. A person who *orders* someone else to carry out some action, or who *asks* the someone else to carry it out, or who *advises* that someone to carry it out, is performing a *directive* illocutionary act. Directive acts are sometimes confused with imperatives, but orders, or commands, are only one kind of directive act. What gives these acts their directive character is their being designed and intended to get the addressee to do or not do something. Directive acts require an addressee. You can't ask someone to pass the salt if there is no someone there.

The third category of illocutionary acts that I will consider here are *commissives*. The person who performs a commissive illocutionary act commits herself to perform, or to refrain from performing, some further act, ordinarily a non-illocutionary act. Promises may be the most well-known kind of commissive act, but if I say to myself, or someone else, "I think I'll get a bottle of beer from the refrigerator in the basement," I have committed myself to act without promising anything to anyone. Like assertive acts, a commissive act may or may not have an addressee, though promises seem to be an exception to this. People make promises to other people. Searle's two other categories of illocutionary acts are *expressives* and *declarations*, but they aren't closely linked to the topics I am discussing in this paper.

2 Locutionary acts

Upon analysis, a relatively straightforward illocutionary act can be recognized to be constituted by performing another language act, a *locutionary act*, with a certain *force*, or in a certain *manner*. The force or manner determines the "job" the speaker is carrying out by performing the locutionary act. (The 'locutionary,' 'illocutionary' terminology is due to Austin, and is explained in Austin, 1965.)

John Searle (1969, 1985) has objected to locutionary acts, either because they don't exist or, if they do exist, because they are simply abstract components of the concrete illocutionary acts that people perform. They are not abstract in the way that numbers, properties, and propositions are abstract, but are abstract by being "what is left" if we (mentally) subtract from an illocutionary act the force which gives that act its distinctive character. Although a locutionary act can be performed, on its own, to illustrate locutionary acts, it is true that ordinarily these acts are abstract components of illocutionary acts. But so what? Assertive locutionary acts are the focus of much logical research, and semantic features of these acts play an important role in determining the logically important features of assertive illocutionary acts.

I use the word 'statement' for the locutionary acts that are used to perform assertive illocutionary acts. These statements are typically performed with sentences or sentential clauses, are appropriately evaluated in terms of truth and falsity, and can, among other things, be asserted or denied or supposed to be true, or to be false. For someone to make a simple factual assertion is for her to produce or perform a statement, accepting that statement for representing or presenting things as they are. In denying a statement, a person is not asserting the negation of the statement being denied, instead, the speaker *rules out* the statement's assertion, because the statement fails to represent or present things as they are.

It is much more common, and idiomatic, to deny that Milwaukee is in Illinois by saying:

(1) Milwaukee is not in Illinois.

than it is to use this rather long-winded (and pompous or pretentious) sentence to make a denial:

(2) I deny that Milwaukee is in Illinois.

Sentence (1) can be used (on different occasions) to perform language acts having different structures, but I think it is most common to use (1) to make what used to be called a *judgment of division*. The 'not' functions as an illocutionary force-indicating expression by separating, or dividing, the act of referring to Milwaukee from the act of predicating 'is in Illinois.' The 'not' is used to rule out the assertion that Milwaukee is in Illinois by blocking the formation of the statement 'Milwaukee is in Illinois.'

While sentence (1) might be used naturally to block the assertion of the statement "Milwaukee is in Illinois" without formulating that statement, the speaker who uses sentence 2 to make her denial explicitly produces the statement whose assertion she rules out.

Positive and negative suppositions occur commonly in arguments by natural deduction. Supposing a statement to be true is like temporarily accepting that statement, and supposing one to be false is like temporarily blocking or impeding the positive supposition of the statement. In deductive arguments especially, it is common both to introduce and to discharge suppositions.

The locutionary acts used to constitute positive directives are (second-person) *plans*, like this:

"Mark, please get up from your seat and close the door."

The plan represents the addressee as carrying out, or performing, the directed action. If the addressee does carry out the directed action, he has *implemented* the plan. But in order to implement a plan, the addressee must *intend* to perform the action involved. If Mark stumbles and accidentally knocks the door shut, he has not implemented the plan "Mark, please close the door."

Just as statements can be asserted or denied, so someone can be directed to implement a plan or to refrain from implementing that plan. The speaker who says to Mark, "Don't open that window" is not directing Mark to implement a negative plan, instead he is trying to block, or *rule out*, Mark's implementing the positive plan of opening the window. While a statement is *satisfied* if it is true, a plan is *satisfied* if it is *implemented*.

For commissives, the locutionary acts are first-person plans like "Michael, I will meet you at the Andrews Theater at 7:45 tonight." This plan represents the speaker as carrying out the action that she commits herself to perform. It is *implemented* if she does meet Michael at the theater at 7:45.

There is an important connection between directive acts and commissive acts. For a speaker's directive act "Mark, please close the door" to be suc-

"I Asked You to Mail that Letter, Not to Burn It"

cessful, several things need to happen: Mark must hear, and understand, the speaker, Mark must agree to implement the speaker's plan, and Mark must close the door *in order* to implement that plan. If Mark makes his agreement explicit, he performs a commissive illocutionary act. Even if Mark agrees without saying anything, or thinking any words, Mark's agreement *commits* him to close the door–in such a case we might regard him as tacitly, or "in effect," performing a commissive act.

3 Assertive locutionary and illocutionary arguments

Locutionary and illocutionary acts occur in arguments of different kinds. An assertive *locutionary argument* is an ordered pair whose first member is a set of statements, the *premisses*, and whose second member is a single statement, the *conclusion*. An assertive locutionary argument is *valid* if the premisses entail the conclusion, and is *logically* valid if the premisses *entail* the conclusion on the basis of the logical forms of premisses and conclusion–in that case, the premisses *imply* the conclusion. Familiar logical systems focus on statements, locutionary arguments, and on semantic features of statements that can be traced to their logical forms.

To (informally) represent an assertive locutionary argument from premisses A, B, C to conclusion D, I will use the following:

$< \{ A, B.C \}, D >$

The ordered-pair notation and the braces are intended to show that the argument which is represented involves sets, and so is an abstraction which a person can represent and evaluate, but is not an argument which a person can make or address to someone else.

An *assertive illocutionary argument* is one that someone does make, by reasoning from assertive illocutionary act premisses to a conclusion of that same kind. Here is a simple example:

> Washington, D.C. is the capital of the United States or else Philadelphia is the capital. Philadelphia isn't the capital. So it must be Washington, D.C. which is the US capital.

If this were a serious argument that someone makes, the premisses and the conclusion would be assertions.

Deductive illocutionary arguments depend on *rational commitment*, which is a person's commitment to do or not do something, or a commitment to remain in a certain state, like that of accepting a given statement.

John T. Kearns

Making a decision to carry out a given action rationally commits a person to carry out that action. Performing some intentional acts can rationally commit a person to perform others. Rational commitment is either immediate or mediate. A person's immediate commitments are evident to her if she gives the matter her attention, but if performing act X_1 will immediately commit Anne to perform act X_2, and performing X_2 will immediately commit her to perform X_3,...., and performing X_{n-1} will immediately commit her to perform X_n, then performing X_1 may only *mediately* commit her to perform X_n. A person's mediate commitments may not be evident to her.

Rational commitment is not some kind of causal necessity. Like most people, I make many decisions which I don't carry out. Sometimes I forget what I decided to do, sometimes I change my mind, and sometimes I am unable to perform the action I decided on. Rational commitment, when recognized, *motivates* a person to act, but it may not carry the day. Honoring this commitment is a requirement of reason, and may or may not be a moral requirement. I can decide to do something like get a bottle of beer from the refrigerator, and then fail to do it, either because I forget what I intended to do, or for some other reason, without being culpable in any way.

Some commitments are conditional, like the commitment to close the upstairs windows in my house if it rains while I am at home, and others, like my commitment to get beer from the refrigerator, are unconditional. With the commitment to close the windows, we might say that I am committed to close the windows *on the condition of it raining while I am at home*. But it is more accurate to say that the commitment is *on the condition of my realizing that it is raining while I am at home*. My commitments can't motivate me to act unless I am aware of them, and, in the conditional case, I also need to be aware that they are "in force."

Coming to accept, or continuing to accept, some statements, and rejecting others, will *inferentially* commit a person to accept further statements, and to reject further statements. Positively or negatively supposing statements will commit a person to suppose others (either positively or negatively). If the person who accepts certain statements, and rejects others, is inferentially committed by this to, say, accept statement A, she has an *assertive inferential commitment*. The *inferential* commitment characteristic of assertive illocutionary acts is not a commitment to carry out reasoning, but is instead a commitment, *when* carrying out deductive reasoning, to make "moves" based on immediate commitments which she recognizes.

"I Asked You to Mail that Letter, Not to Burn It"

If asserting A inferentially commits a person to assert B, and supposing A to be the case inferentially commits a person to suppose B, it does not follow that supposing A will inferentially commit her to assert B. An assertion is stronger than a supposition, and can commit the arguer to assert or deny further statements, while suppositions can only commit the arguer to make further suppositions. Inferential commitment is an epistemic feature.

A person can simply recognize her immediate commitments, and can, in principle anyway, come to recognize her mediate commitments. On investigation, some commitments turn out to be based on semantic features like entailment or implication. If statement A entails B, and their connection is easy to grasp, this may help us to understand why asserting or supposing A will inferentially commit someone to assert or suppose B. But not all inferential commitments are based on entailment or implication. If I assert "It is raining," this will commit me to assert "I believe that it is raining," but the statement "It is raining" does not entail the statement that I believe this.

Although I accept many statements and reject many others, I am uninterested in exploring most of the "commitment consequences" of these beliefs and disbeliefs. For example, if I accept the statement "Today is Thursday," I will be inferentially committed to accept the statement "Either today is Thursday or it is now snowing in Beijing." However, I would have no interest in this consequence, and would not be committed either to consider or accept it. Still, if the matter somehow came up, and I took some interest in the issue of whether the statement was true, I would be committed to accept that either today is Thursday or it is now snowing in Beijing. It would be irrational to accept that today is Thursday, but refuse to accept the disjunction.

An assertive illocutionary argument is *deductively correct* if it is simple and the premiss acts rationally and inferentially *commit* the arguer to perform the conclusion act, while if it is complex, all component arguments must be deductively correct and the initial undischarged premiss acts must inferentially commit the arguer to perform the conclusion act. (Logical theories for assertive illocutionary arguments are presented and discussed in Kearns (1997, 2000, 2006, 2007). The understanding of illocutionary arguments, and the theories themselves, are better in the later papers than in the earlier ones.)

4 Directive arguments

In addition to assertive arguments, there are also directive and commissive arguments. A *directive illocutionary argument* has a conclusion which is a directive act and premises which give the addressee reasons to implement the conclusion's plan. But the premises of a directive illocutionary argument are not themselves directive acts. In this directive illocutionary argument:

> Tara, your mother will be offended, and hurt, if you don't go home for the holidays. So you must visit your family for Christmas.

the premiss is an assertion. This argument doesn't seem properly regarded as deductive, for the speaker is trying to get the addressee to, first, *decide* to visit her family, and, subsequently, to make the visit. The truth of the premiss isn't sufficient to commit the addressee to implement the conclusion.

But this directive argument:

> Kevin, you have promised to pick up Max from soccer practice. His practice ends in 15 minutes, and it takes that long to get to the field from here. So please go now to get Max.

is intended to make Kevin realize that he has *already* committed himself to implement the conclusion's plan. It seems reasonable to regard this as a deductively correct argument.

A *directive locutionary argument* is an ordered pair whose first member is a set of premises which are either statements or plans addressed to a single addressee, and whose second member, the conclusion, is a plan with the same addressee as the premiss plans. We could represent such an argument like this:

> $<$ {Kevin, please pick up Max after soccer practice; Max's soccer practice ends in 15 minutes; It takes 15 minutes to drive from here to the practice field.}, Kevin, please go now to pick up Max from practice. $>$

This argument is valid, because (roughly) any way of satisfying the premisses will satisfy the conclusion, but this directive *locutionary* argument is not closely related to a directive *illocutionary* argument. The premisses, if

"I Asked You to Mail that Letter, Not to Burn It"

actually addressed, or spoken, to Kevin, give Kevin no reason to implement the conclusion.

There are puzzles concerning disjunction that are sometimes raised in connection with directive acts and arguments, and these puzzles might make someone dubious about the possibility of developing a logical theory which accommodates directives. If Vladimir makes a request to Jaroslav by saying "please mail this letter," handing him a sealed envelope with postage attached, he has asked, and directed, Jaroslav to place the stamped envelope in the mail. And if he has asked Jaroslav to mail the letter, then, clearly, he has either asked Jaroslav to mail the letter or he has asked Jaroslav to burn the letter. But we all recognize that Vladimir has not asked Jaroslav to either mail or burn the letter.

This is a kind of situation which many people have found puzzling. Why hasn't Vladimir made a disjunctive request? If we change our story a little, and have Vladimir handing Jaroslav an envelope with no postage attached, saying "Please put stamps on this so that it can be mailed," and then later saying to Jaroslav, "Please mail that envelope I handed you," it seems OK to say that Vladimir asked Jaroslav to put postage on the envelope and to mail it, although Vladimir did not make one "conjunctive" request. Why is *conjunction introduction* OK here, when *disjunction introduction* is not OK?

Our situating these stories in a speech-act framework allows us to pretty much "erase" their puzzling features. Assertive locutionary acts, or statements, have truth conditions which can be satisfied or not. We define the relations of entailment and implication linking statements in terms of the satisfaction conditions of statements. Plans which are the locutionary components of directive acts have implementation conditions rather than truth conditions, but these conditions can be satisfied or not. And we can define relations of directive entailment and implication linking second-person plans which have the same addressee and are intended for the same occasion. For example, suppose that "Michael, do F" and "Michael, do G" are such plans. Then the "do F" plan entails the "do G" plan iff any way in which Michael implements the "do F" plan will also implement the "do G" plan. In the right setting, the plan:

Michael, get up from your seat and shut the door.

will directively entail this plan: Michael, get up from your seat.
Michael cannot implement the first plan without also implementing the second.

If we were concerned with a real event concerning Vladimir and Jaroslav, an inference from this assertion:

> Vladimir asked Jaroslav to mail the letter.

to this one:

> Vladimir asked Jaroslav to mail the letter or he asked Jaroslav to burn the letter.

would be an assertive illocutionary inference which exemplifies the principle *disjunction introduction*, and it would be "deductively correct." An inference from:

> Vladimir asked Jaroslav to mail the letter.

to:

> Vladimir asked Jaroslav to mail the letter or burn it.

would also be an assertive illocutionary inference, but it does not exemplify the principle *disjunction introduction*, and it isn't correct.

For this plan:

> Jaroslav, please mail this letter.

does not directively entail:

> Jaroslav, please mail this letter or burn it.

To implement the simpler plan, all Jaroslav needs to do is mail the letter, while to implement the disjunctive plan, Jaroslav needs to do two things:

> (1) consider the plan's disjuncts and select one
>
> (2) implement that disjunct.

Jaroslav can clearly implement the simpler plan without doing those two things.

All it takes to satisfy a disjunctive statement '*A or B*' is that one of the disjuncts be true, but it isn't sufficient to satisfy a disjunctive plan '*do F or do G*' that one of the disjuncts be implemented. The addressee must *intend* to *do F or do G*, and this requires him to consider both disjuncts when he chooses one to implement.

This is the right answer to the problem concerning disjunction and *disjunction introduction* for directives, but it calls into question what I said about *conjunction introduction*. For it isn't in general the case that if Mark has been asked to *do F* and has also been asked to *do G*, then he has been asked to *do F and do G*. Just as Mark would need a disjunctive intention to implement a disjunctive plan, so he should have a conjunctive intention to implement a conjunctive plan. But he wouldn't have that if he wasn't asked to carry out a conjunctive plan.

So, in the earlier story, was Jaroslav asked to put postage on the letter and mail it? The first request that Vladimir made was only for Jaroslav to put postage on the envelope. The second request was a request for Jaroslav to mail the letter. We aren't told whether Jaroslav had already put postage on the letter. But you can't really mail a letter if it has no postage attached. You can put the letter in a mail box, but it won't be delivered. In case Jaroslav had not already put postage on the envelope, the second request would cover both putting on postage and dropping the stamped letter in the mail box. In the example above, it seems OK to me to say that Jaroslav was asked to do both of two things, but that assertion cannot be correctly inferred by *conjunction introduction* from two premiss assertions, one for the postage and one for the mailing.

5 Articulating a conceptual framework and developing logical theories

In Kearns (2015), I discuss assertive, directive, and commissive acts, both locutionary and illocutionary, as well as locutionary and illocutionary arguments for the three categories of assertive, directive, and commissive acts. I explore the relations between the different kind of acts and arguments, primarily in order to call attention to the different kinds of arguments for which we can develop distinctive logical theories. I described what I was doing in that paper as *articulating a conceptual framework* for the various acts and arguments. That framework "*lays out a logical landscape that can be investigated and explored by a variety of logical theories.*"

The present paper amends and corrects that conceptual framework. In the earlier paper, I somewhat misunderstood directive locutionary and illocutionary acts, failing, in particular, to understand the implementation conditions of disjunctive plans. I have fixed that misunderstanding in this present paper. The correction shows that the treatment of directive acts and arguments can be less complicated than I previously thought, but it also reveals a new respect in which directive acts and arguments differ from assertive acts and arguments.

Although I have further articulated, and corrected, the conceptual framework I outlined in Kearns (2015), I haven't developed and presented a logical theory for either directive locutionary arguments or directive illocutionary arguments. Developing such theories is an interesting project, and might even turn out to be important. But life is short and art, even the art of logic, is long. I hope that someone else will be motivated to use the framework I have articulated as a guide for developing directive logical theories.

References

Austin, J. (1965). *How to Do Things with Words*. New York: Oxford University Press.
Kearns, J. T. (1997). Propositional logic of supposition and assertion. *Notre Dame Journal of Formal Logic*, *38*, 325–349.
Kearns, J. T. (2000). An illocutionary logical explanation of the surprise execution. *History and Philosophy of Logic*, *20*, 195–214.
Kearns, J. T. (2006). Conditional assertion, denial, and supposition as illocutionary acts. *Linguistics and Philosophy*, *29*, 455–485.
Kearns, J. T. (2007). An illocutionary logical explanation of the liar paradox. *History and Philosophy of Logic*, *28*, 31–66.
Kearns, J. T. (2015). The larger logical picture. In P. Arazim & M. Dančák (Eds.), *The Logica Yearbook* (pp. 107–116). London: College Publications.
Searle, J. R. (1969). *Speech Acts: An Essay in the Philosophy of Language*. London: Cambridge University Press.
Searle, J. R. (1985). *Expression and Meaning: Studies in the Theory of Speech Acts*. Cambridge: Cambridge University Press.

John T. Kearns
University at Buffalo, SUNY,
Department of Philosophy and Center for Cognitive Science
USA
E-mail: kearns@buffalo.edu

The Logical Form of Identity Criteria

ANSTEN KLEV[1]

Abstract: It is argued here that criteria of identity do not have the form of predicate-logical formulae. The conclusion is drawn that Hume's Principle cannot serve as a criterion of identity for the concept of cardinal number. The way criteria of identity are formulated in Martin-Löf's type theory is presented as an alternative that is not affected by the argument.

Keywords: Criteria of identity, Logical form, Neo-Logicism, Type theory

1 Sortal concepts

In accordance with established philosophical terminology let us say that a sortal concept is a concept with which there are associated so-called criteria of application and identity (e.g. Grandy, 2016). The criterion of application associated with a sortal concept C determines what it is for a thing to fall under C. The criterion of identity associated with C determines what it is for things falling under C to be the same object. For example, one could propose as the criterion of application associated with the concept of a truth-value that x is a truth-value if it evaluates to either the True ($\top$) or the False ($\bot$); and as the criterion of identity that truth-values x and y (perhaps given to us through certain complicated boolean expressions) are identical if either both evaluate to the True or both evaluate to the False. This example is especially simple since the criterion of application is here given by a list of the 'canonical' objects falling under C; and it may be peculiar in that objects falling under C are objects of which it makes sense to say that they evaluate to such and such. But it serves to show that the notions of criteria of application and identity are not vacuous.

The criterion of application associated with C may well be taken to provide an answer to the question of what a thing falling under C is. What is

---

[1] Readers of Sundholm (1999) will see my indebtedness to him. I am grateful to Bob Hale for his comments on my talk at Hejnice and for our stimulating discussions there on logic, philosophy, and music. This paper is dedicated to his memory.

While writing this paper, the author has been supported by grant nr. 17-18344Y from the Czech Science Foundation, GAČR.

Socrates? A man. What is $\top \supset ((\bot \vee \bot) \supset \bot)$? A truth-value. Thus the notion of a sortal concept is connected with the Aristotelian–Porphyrian notion of a *genus*, since a *genus* was described by both Aristotle and Porphyry (following Aristotle) as a general term which it is appropriate to propose in answer to the question of what something is.[2] This connotation of the word 'sortal' must have been clear to Locke, who first coined this adjective, since he Englished the Greek *genos* as 'sort' and associated with it his doctrine of nominal essence (cf. Locke, 1690, III.iii).

That the concept C is associated with a criterion of identity means that we may speak in the plural about the C's, a number of C's, five C's, etc. In particular, the things falling under C are clearly distinguishable from each other. The notion of a thing falling under a sortal concept thus provides us with a natural notion of object: whatever falls under a sortal concept has a 'nominal essence' and is a clearly circumscribed part of the world. (This is not to say that there might not be other, quite different, accounts of the notion of object.) Since also concepts under which functions fall may be given criteria of application and identity, one sees that this way of employing the word 'object' was not Frege's, since he contrasted object and function (Frege, 1893, §§ 1–2). Namely, we may say that a function f from A to B is a thing such that i) whenever a is an object falling under A, then $f(a)$, called the result of applying f to a, or the value of f at a, is an object falling under B; and ii) whenever a and a' are identical objects falling under A, then $f(a)$ and $f(a')$ are identical objects falling under B. The functions f and g from A to B are identical if $f(a)$ is identical with $g(a)$ whenever a is an object falling under A. Provided A and B themselves are sortal concepts, these explanations constitute criteria of application and identity for the concept of a function from A to B.

The criterion of identity of one sortal concept may not coincide with that of another, as was in effect noted by Locke (1690, II.xxvii.7) and assumed by Frege in the considerations leading up to his definition of the concept of number (*Anzahl*) in § 69 of the *Grundlagen der Arithmetik* (1884) as well as by Cantor in his first definition of the concept of set.[3] It must be emphasized, however, that the relativity of identity criteria to sortal concepts does not entail the doctrine of relative identity, defended by Geach (1972, pp. 238–247) and Griffin (1977). According to this doctrine the notion of identity as such is relative to sortal concepts, so that a and b may be the same C but

---

[2] For Aristotle, see *Topics* 102a32–35, and for Porphyry, see Barnes (2003, p. 4).
[3] Cf. Cantor (1882, pp. 113–114).

fail to be the same D, where C and D are sortal concepts under which both a and b fall. The doctrine of relative identity follows only on the assumption that—and is in effect tantamount to the claim that—a thing may fall under two sortal concepts associated with different criteria of identity.

2 Neo-Fregeanism

According to the Neo-Fregeanism defended by Hale and Wright (e.g. 2001), *cardinal number* is a sortal concept whose criteria of application and identity both flow from what, following Boolos (1987, p. 186), is called Hume's Principle:

$$\#F = \#G \Longleftrightarrow F \sim G$$

Here # is a second-order function symbol which, when applied to a first-order predicate symbol F, yields an individual term $\#F$; and $F \sim G$ is a formula (definable in pure second-order logic) saying that there is a 1–1 relation between the F's and the G's. More precisely, the criterion of identity associated with the concept of cardinal number is to be Hume's Principle itself. The postulation of this formula as an axiom is, moreover, to serve as a definition, albeit a non-explicit definition, of the second-order function #. The criterion of application of the concept of cardinal number is then to be given as

$$\text{Card}(n) \Longleftrightarrow \exists F(n = \#F)$$

A technical result appealed to in support of this conception of cardinal number is what, also following Boolos (1990, p. 209), is called Frege's Theorem: the second-order Dedekind–Peano axioms of arithmetic are derivable in second-order logic extended with Hume's Principle (cf. e.g. Wright, 1983, sec. xix). This latter theory, called Frege Arithmetic, is moreover consistent relative to the former (Boolos, 1987).

The Neo-Fregean claim that Hume's Principle supplies the concept of cardinal number with criteria of application and identity goes beyond this technical result and can be challenged on philosophical grounds. The so-called Julius Caesar objection, raised already by Frege himself and much discussed in the literature,[4] questions whether Hume's Principle provides the required criterion of application: the Principle fails, namely, to determine whether the famous Roman general is (or, was) a number. More precisely, for n not of the form $\#G$, the only way an equality of the form $n = \#F$

---

[4]E.g. Frege (1884, § 65), Wright (1983, sec. xix), and Hale and Wright (2001, ch. 14).

can be established or refuted inside Frege Arithmetic is through a sequence of explicit definitions showing $n = \#G$ for some G; but it is clear that Julius Caesar is not definitionally equal to any $\#G$.

In this paper I wish to challenge the apparently less controversial claim that Hume's Principle provides a criterion of identity for the concept of cardinal number. In fact, the argument will have a much wider scope. I shall, namely, argue that the logical form of an identity criterion cannot be that of an ordinary predicate-logical formula. In particular, the logical form of the criterion of identity associated with the sortal concept C cannot be that of a formula

$$t[\bar{x}] = t'[\bar{y}] \iff \psi[\bar{x}, \bar{y}]$$

where t and t' are terms, ψ a formula, and square brackets indicate variable-occurrence; the sequence of variables $\bar{x}$ may include variables ranging over higher types. Since Hume's Principle has this form, it follows that it cannot serve as a criterion of identity for the concept of cardinal number. Although the argument is quite general, I have decided to concentrate on Hume's Principle both for concreteness and so as to point immediately to an important corollary of the general argument. After this mainly negative argument, I shall illustrate, by means of reference to Martin-Löf's type theory, how criteria of identity can be formulated.

3 The general argument

As is well-known, an important episode in Frege's logical revolution was his introduction of the function–argument analysis of judgeable contents (I am using this happy term without commitment to the specifics of the *Begriffsschrift* (1879), where it properly belongs within Frege-scholarship). Any judgeable content is to be analyzed as: function applied to one or more arguments. This analysis leads naturally to a type-theoretical treatment of logical syntax, according to which any component of a judgeable content is taken to belong to a certain type in a type hierarchy. In the well-known case of the simple type hierarchy (e.g. Church, 1940), already implicit in Frege's *Grundgesetze* (1893), there is a type ι of individuals, a type o of truth-values, and function types built up inductively from these. A n-ary relation of individuals is then treated as a n-ary function from ι into o; a binary propositional connective as a binary function from o into o; and a first-order quantifier is treated as a function taking functions from ι into o as arguments and truth-values as values. If the notion of a universal domain of individuals

Identity Criteria

is rejected as unintelligible and replaced by an indefinite range of smaller such domains, as in Martin-Löf's type theory, then a hierarchy of dependent types is called for (cf. Aspinall & Hofmann, 2005). The adaptation of type-theoretical grammar to natural languages, as in categorial grammar (cf. Morrill, 2011), requires refinements of other kinds, but the basic structure of function–argument remains essentially as before.

A type-theoretical analysis of the full language of predicate logic, in particular of the quantifiers, is not presupposed by the argument to be presented here. Only for the atomic formulae of the language do we assume such an analysis: an atomic formula is obtained by the application of a function to a suitable number of arguments. This analysis of atomic formulae, which is as Fregean as anything can be in logic (e.g. Frege, 1891), is standard and can be found in classical logic books such as Hilbert and Ackermann (1928), Hilbert and Bernays (1934), Kleene (1952), and Church (1956). It motivates the use of the label 'functional calculus' in the first and last works on this list for what the other two call 'predicate calculus' (both explaining 'predicate' as propositional function).

Before presenting the argument I should remark that I am assuming predicate logic to be a meaningful symbolism: it is a medium for the expression of thought, and not a system of structured objects, about which we may form certain thoughts, but with which no such thoughts can be expressed (cf. Sundholm, 2002). This assumption accords well with Neo-Fregeanism, since Hume's Principle is written in this symbolism and taken to be meaningful. I will therefore speak not only of 'formulae' but also of 'propositions', not only of 'predicates' but also of 'propositional functions'. One may regard 'proposition' and 'propositional function' as names of categories of meaningful expressions, or—what is commoner—as names of more purely ontological, less linguistic, categories, residents of a Fregean realm of sense, say. When concentrating on the formal aspect, or surface aspect, of the expressions of predicate logic I shall use '(well-formed) formula' and 'predicate'.

Also truth-values will be appealed to below. When doing so I shall think of propositions as ways of specifying truth-values. As Church (1956, p. 28) notes, in this case a propositional function is really a truth-value function, since the values such a function takes—in the usual sense of value of a function—are truth-values and not propositions. If truth-values are not appealed to in the semantics of predicate logic, as in the semantics of constructive logic, then propositions may be regarded as the values, in the usual sense, of propositional functions. Now for the argument.

Ansten Klev

It happens often in mathematics that one must show a certain function to be well-defined. What is then meant by the well-definedness of a function f is that it satisfies the following congruence condition:

$$x = y \Longrightarrow f(x) = f(y) \qquad \text{(Cong)}$$

For identical arguments the function must yield identical values. Thus, a 'function' f on the rational numbers defined by

$$f(\frac{m}{n}) := m + n$$

is not well-defined, and so in effect, not a function at all. From this example it is clear that well-definedness is a requirement of extensionality: the result of applying the function to an argument must be independent of the 'mode of givenness' of the argument.

Since propositional functions are functions, it is meaningful to ask whether a given propositional function is well-defined. If truth-values are assumed, and 'propositional function' means truth-value function, the well-definedness of a propositional function amounts to a familiar Fregean tenet (e.g. Frege, 1892): the truth-value of a sentence S is not altered by the replacement of a constituent c of S by an expression c' with the same reference as c. Since sameness of truth-value is captured by the relation of material equivalence, the condition of well-definedness for propositional functions can thus be phrased as

$$x = y \Longrightarrow (P(x) \Longleftrightarrow P(y)) \qquad \text{(Cong*)}$$

If truth-values are not appealed to, and the values of propositional functions are propositions, then a notion of identity of propositions is called for. Here there are several options, but for current purposes it is enough to assume the very weak notion of identity captured by material equivalence, whence well-definedness is again expressed by (Cong*). An example of a 'propositional function' not satisfying (Cong*) is the 'function' W defined on the rational numbers as

$$W(\frac{m}{n}) := m \text{ is prime} \wedge n > 5$$

Applied to $\frac{3}{6}$ this W would yield a true proposition (the True) and applied to $\frac{4}{8}$ a false proposition (the False), even though $\frac{3}{6} = \frac{4}{8}$.

From (Cong) it is clear that the question of whether a function f is well-defined presupposes that the domain of f is equipped with identity: we

Identity Criteria

cannot make sense of this question unless we know what it is for admissible arguments to f to be identical. The question of the well-definedness of a propositional function, in particular, presupposes that we know what it is for elements of the underlying domain to be identical. But, what it is for elements of the underlying domain to be identical is just what is determined by a criterion of identity. The question of the well-definedness of a propositional function therefore presupposes that a criterion of identity is already associated with the underlying domain.

The left-hand side of Hume's Principle is an atomic formula,

$$\#F = \#G$$

whose outermost structure is: binary predicate applied to two individual terms as arguments. In particular, identity here appears as a propositional function. The question of well-definedness can be raised for this propositional function just as it can for any propositional function, or purported propositional function. And, as we have just seen, this question presupposes that a criterion of identity is associated with the underlying domain. It is natural to require of a well-formed formula that it be built up using only well-defined predicates. A formula involving the W above, for instance, is not well-formed, since it contains what purports by its form to be a propositional function, but which in fact is no such thing. Hence, we can be certain of the well-formedness of the left-hand side of Hume's Principle, and consequently of the whole Principle, only if we know a criterion of identity associated with the underlying domain. It should be clear, then, that this criterion of identity cannot be provided by Hume's Principle.

The argument is, as already noted, general. A criterion of identity cannot be given in the form of

$$t[\bar{x}] = t'[\bar{y}] \iff \psi[\bar{x}, \bar{y}] \qquad \text{(Id-Crit)}$$

since here identity, =, appears as a propositional function and so presupposes for its well-definedness an underlying notion of identity.

One can reach the same conclusion from two other, slightly different, routes. Any function must have a domain, often called its domain of definition. But, in order for this domain to be the domain of a function, there must be a criterion of identity associated with it, since being a function f on the domain A requires that whenever a and a' are identical arguments to f, then $f(a)$ and $f(a')$ are identical values. In fact, we saw above that for *function from A to B* to be a sortal concept it must be assumed that A and B

themselves are sortal concepts, in particular that a criterion of identity is associated with each of them. A propositional function, being a function, must thus have a domain, in this case often called a range of significance, and this domain must be equipped with a notion of identity. A criterion of identity cannot, therefore, be given in the form of (Id-Crit), where = appears as a propositional function, since here, by way of the domain of the =-function, a notion of identity is already assumed.

The other alternative route goes via the notion of an 'underlying domain' appealed to at several places above. Each predicate has an underlying domain, namely a range of significance; as does each variable, free or bound, namely the domain over which it ranges, and which, when the variable is bound by a quantifier, is also a domain of quantification. But of any such domain, in particular of any domain of quantification, it seems reasonable to require that it be equipped with identity. If I have proved that a certain predicate P is true of 0 and of every successor $s(n)$, then I can infer that P is true of every natural number, since any natural number is identical either to 0 or to a successor. It is thus not then open to someone to object, on the grounds that 11^{13} is not of successor form, that I have not proved that P is true of 11^{13}. In proving that a universally quantified proposition is true, we of course need to be sure that we have covered all the cases, and as this example of mathematical induction shows, the notion of all cases would seem to presuppose a notion of identity in the underlying domain. More generally, a notion of identity would seem to be presupposed by the conception of the underlying domain as a domain of clearly circumscribed objects.[5]

4 Objections and replies

An obvious response to this family of arguments is the objection that = as it appears in (Id-Crit) is not a propositional function, or more generally, not a function. To carry weight, this objection should be accompanied by a clear conception of the syntax of predicate logic that does not rely on the notion of function. The only such conception of which I am aware treats predicate symbols as syncategorematic: by itself the predicate P signifies nothing, but when appended to some individual term a, we get a proposition Pa (depending on a variable assignment in case a contains free variables). This is in effect the conception of propositional connectives assumed by Church

---

[5]Notice that these considerations on the relation between identity and quantification are quite different from those, involving variable recurrence, attacked by Wehmeier (2017).

Identity Criteria

(1956, § 5), which can readily be extended to predicates. It can, moreover, be extended to the quantifiers by introducing in addition the notion of variable-binding (ibid. § 6). We thus have a conception of the syntax of predicate logic according to which individual terms and well-formed formulae are categorematic and all other signs syncategorematic. However, even when treated as syncategorematic, predicates are still subject to a requirement of congruence: the truth-value of an atomic formula must still remain the same when an individual term occurring in it is replaced by an individual term with the same reference. Hence, also on this conception of logical syntax must a notion of identity be presupposed for the domain of reference of the individual terms.

Another response is to insist that function–argument structure is found only at the formal level of syntax. A predicate symbol, treated as a formal object, is a function, which when applied to a suitable number of suitable arguments, themselves treated as formal objects, is another formal object, called a well-formed formula.[6] But the predicate, although it is a function, does not refer to a function. This is clear from the standard model-theoretic semantics of predicate logic, where predicates are interpreted as sets of ordered pairs and not as functions. The arguments above thus rest on a confusion of syntax and semantics: we have functions on the level of syntax, but not on the level of semantics.

The objection fails, however, since model-theoretic semantics, provided we want to regard it as a semantics and not as a purely mathematical construction, presupposes that the (predicate-logical) language of set theory is meaningful. Consider the clause in this semantics for an atomic formula Pa. With an interpretation function $\mathfrak{J}$, the clause is

$$\mathfrak{J}(a) \in \mathfrak{J}(P)$$

This must be regarded as a meaningful formula of set theory, since it is meant to explain the meaning of Pa (or, to be more precise, the meaning of $\mathfrak{J} \models Pa$). The arguments of the previous section certainly apply to the language of set theory. They therefore show that an interpretation in the model-theoretic sense of Hume's Principle presupposes that the domain of the interpretation has been equipped with identity. It is known that the (first-order) domain of the interpretation of Hume's Principle must be assumed to include the interpretations of the individual terms $\#F$, since otherwise one cannot establish that there are infinitely many cardinal numbers (cf. Heck,

---

[6]The term 'formal object' is taken from Curry (1963, ch.2).

1997). Hence, the domain of the interpretation must both be equipped with identity and include the interpretation of the terms $\#F$, to wit, the cardinal numbers. Hume's Principle thus taken therefore presupposes that a criterion of identity for cardinal numbers is already present.

A final response I shall consider that may be open to the Neo-Fregean is the objection that Hume's Principle as such does not presuppose the conceptual architecture of predicate logic. Granted, the Principle is usually written in the language of predicate logic, but that is only because this allows for a compact, convenient formulation of it. Its being so written is no precondition for its serving as an explanation of the concept of cardinal number. It can fulfil this task also when written, for instance, in semi-technical English: the number of objects falling under some concept F is equal to the number of objects falling under some concept G if and only if there is a one-to-one correspondence between the objects falling under F and those falling under G.

This response drives a wedge between the Hume's Principle that is meant to explain the concept of cardinal number and the Hume's Principle that is an axiom of Frege Arithmetic, since the latter certainly is a formula of predicate logic. But if these two versions of Hume's Principle are thus separated, then it would seem that Frege's Theorem can no longer be appealed to in support of Neo-Fregeanism, since the Hume's Principle mentioned in Frege's Theorem is then not what, according to Neo-Fregeanism, is meant to provide the sortal concept of cardinal number with criteria of application and identity. It is indeed questionable whether the response gets off the ground at all, since even the English formulation of Hume's Principle avails itself of the Fregean vocabulary of concepts and objects and would thereby seem to be committed to the conceptual architecture of predicate logic. It seems, moreover, that Neo-Fregeans themselves take Hume's Principle to presuppose the resources of predicate logic. For, when Wright (1998) explains how a person is to attain an understanding of the concept of cardinal number with the help of this Principle, he assumes that the person 'has mastered an appropriate higher-order logic, in which the *definiens* of $N^=$ [i.e. of Hume's Principle] can be formulated' (ibid. p. 248).

5 The proper treatment of identity

The arguments above, to the extent that they are successful, show that criteria of identity do not have the form of predicate-logical formulae; taking

Identity Criteria

such formulae to signify propositions, the arguments thus show that criteria of identity are not propositional in form. They do not show that the propositional function of identity as such is defective. They show only that this function is not fit for the task of expressing identity criteria. There might well be other tasks for which it is fit. Indeed, the use of identity in ordinary predicate-logical languages suggests that the propositional function of identity routinely enters into the composition of the propositions expressed in such languages, as for instance into the law of commutativity, $(\forall x, y \in \mathbb{N})(x+y = y+x)$. And the arguments certainly do not show that this use of the identity function is somehow flawed.

We seem to be led to a distinction drawn by Sundholm (1999, p. 26) between criterial identity and propositional identity. Accepting this distinction we are no longer conceiving of identity as one, simple relation, but we are recognizing at least two relations of identity. We certainly expect that if two things are criterially identical, then they are also propositionally identical; and we might expect that the inference in the other direction is valid as well (but see below). Thus we may expect to have extensionally the same relation. But intensionally (or is this a case where one should say 'hyperintensionally'?) the relations differ. They thus differ since, as the arguments above show, criterial identity is conceptually prior to propositional identity: before we can speak meaningfully of a propositional *function* of identity, a criterion of identity must already be in place.

By saying that criterial identity differs from propositional identity, we are in effect also saying, or repeating, that criteria of identity are not to be formulated using propositional identity. The question therefore arises how criteria of identity are to be expressed. An answer that readily suggests itself is that we need to introduce a new form of statement for this purpose: an extension of the syntax of predicate logic is required if we wish to express criteria of identity in a predicate-logical language, since the ordinary predicate of identity in this language expresses propositional identity.

It may be useful to see an example of how such an extension can be implemented. In Martin-Löf's type theory a novel form of identity statement is part and parcel of the basic syntax and is used, among other things, in the formulation of identity criteria. The form of statement is

$$a = b : \alpha,$$

to be read as 'a and b are identical objects of type α', or also 'identical α's' or 'the same α's'. In the terminology of type theory this form of statement is called an identity *judgement*. Such a judgement differs from the identity *proposition* $a =_A b$ that can be formed whenever A is a type of individuals and a and b objects of type A. The connectives and the quantifiers operate on propositions, but not on judgements. Moreover, the premises and conclusions of inferences in the theory are judgements, whereas propositions feature only as parts of judgements. In particular, given a proposition $a =_A b$, there is a form of judgement

$$a =_A b \text{ true}$$

asserting that the proposition $a =_A b$ is true. This is an instance of the general form of judgement 'A true', where A is any proposition, and which may usefully be compared to the form of judgement

$$\vdash A$$

in Frege's logic, involving the so-called judgement stroke.

The conceptual priority of the identity judgement $a = b : A$ over the identity proposition $a =_A b$ is very clear in type theory. The formation of the proposition $a =_A b$, namely, presupposes that A is a type; but for A to be a type, there must be associated with it a criterion of identity, and this is formulated using judgements of the form $a = b : A$. The priority of the domain of definition of a propositional function over the function itself is assumed also in the standard model-theoretic semantics of predicate logic: one or (as in many-sorted logic) more domains are given first of all, and the predicate symbols are thereafter interpreted over these. In type theory, however, the domains are not given by some outside theory (axiomatic set theory, say), but defined inside the theory itself, in effect through the statement of criteria of application and identity.

In particular, the identity criterion of a type A of individuals involves two components. One component is specific to the type A and determines what it is for so-called canonical elements of A to be identical. This component typically consists of certain stipulations similar to the introduction rules of natural deduction. In the case of the natural numbers, $\mathbb{N}$, for instance, the stipulations look as follows:

$$0 = 0 : \mathbb{N} \qquad \frac{m = n : \mathbb{N}}{s(m) = s(n) : \mathbb{N}}$$

Thus, it is stipulated that 0 is the same canonical $\mathbb{N}$ as 0 and that, if m and n are the same $\mathbb{N}$ (they need not be canonical), then $s(m)$ and $s(n)$ are identical canonical $\mathbb{N}$'s. The second part of the identity criterion is common

Identity Criteria

to all types of individuals and says that a and b are the same A provided they evaluate to identical canonical elements of A. The notion of evaluation to canonical form appealed to here is one of the primitive notions involved in the intended interpretation of Martin-Löf's type theory.[7] The appeal to this notion reflects the fact that this type theory is intended as a framework for constructive mathematics.

The rules of Martin-Löf's type theory allow one to infer the judgement $a =_A b$ true from the judgement $a = b : A$. The inference in the other direction is, however, a more complicated matter. One can, as in the version of the theory presented in (Martin-Löf, 1984), postulate that inference as a primitive rule. This rule is justified by the intended interpretation, but it does not accord well with the idea—which will not be discussed here—that $a = b : A$ is also to mean that a and b are definitionally equal.[8] The rules pertaining to identity found in (Martin-Löf, 1975b) may therefore seem preferable. Using these rules, however, one cannot in general infer $a = b : A$ from $a =_A b$ true. It is a metamathematical result (namely, a corollary of normalization) that if there is a closed derivation of $a =_A b$ true, then there is also a derivation of $a = b : A$.[9] But if a and b are allowed to be open terms, then this inference is no longer valid. For instance, although one can derive

$$x+y =_\mathbb{N} y+x \text{ true}$$

for variables $x, y : \mathbb{N}$, one cannot derive $x+y = y+x : \mathbb{N}$. In this version of Martin-Löf's type theory, therefore, propositional identity is strictly weaker than criterial identity: although the inference from a criterial identity to the corresponding propositional identity is always admissible, the inference in the other direction is not always admissible.

References

Aspinall, D., & Hofmann, M. (2005). Dependent types. In B. C. Pierce (Ed.), *Advanced Topics in Types and Programming Languages* (pp. 45–86). Cambridge, MA: MIT Press.

Barnes, J. (2003). *Porphyry. Introduction. Translated with an Introduction and Commentary*. Oxford: Oxford University Press.

---

[7] For an account of the intended interpretation, see Martin-Löf (1984) or Klev (forthcomming).
[8] On the notion of definitional equality, see Martin-Löf (1975a).
[9] See Martin-Löf (1975b, Theorem 3.14).

Boolos, G. (1987). The consistency of Frege's *Foundations of Arithmetic*. In J. J. Thomson (Ed.), *On Being and Saying: Essays in Honor of Richard Cartwright* (pp. 3–20). Cambridge: MIT Press. (Cited from Boolos, 1998)

Boolos, G. (1990). The standard equality of numbers. In G. Boolos (Ed.), *Meaning and Method: Essays in honor of Hilary Putnam* (pp. 261–267). Cambridge University Press. (Cited from Boolos, 1998)

Boolos, G. (1998). *Logic, Logic, and Logic*. Cambridge: Harvard University Press.

Cantor, G. (1882). Ueber unendliche, lineare Punktmannichfaltigkeiten. *Mathematische Annalen*, 20, 113–121.

Church, A. (1940). A formulation of the simple theory of types. *Journal of Symbolic Logic*, 5, 56–68.

Church, A. (1956). *Introduction to Mathematical Logic*. Princeton: Princeton University Press.

Curry, H. B. (1963). *Foundations of Mathematical Logic*. New York: McGraw-Hill.

Frege, G. (1879). *Begriffsschrift*. Halle: Louis Nebert.

Frege, G. (1884). *Grundlagen der Arithmetik*. Breslau: Verlag von Wilhelm Koebner.

Frege, G. (1891). *Funktion und Begriff*. Jena: Hermann Pohle.

Frege, G. (1892). Über Sinn und Bedeutung. *Zeitschrift für Philosophie und philosophische Kritik*, *NF 100*, 25–50.

Frege, G. (1893). *Grundgesetze der Arithmetik I*. Jena: Hermann Pohle.

Geach, P. T. (1972). *Logic Matters*. Oxford: Blackwell.

Grandy, R. (2016). Sortals. In E. N. Zalta (Ed.), *The Stanford Encyclopedia of Philosophy*. Standford University.

Griffin, N. (1977). *Relative Identity*. Oxford: Clarendon Press.

Hale, B., & Wright, C. (2001). *The Reason's Proper Study*. Oxford: Oxford University Press.

Heck, R. G. (1997). The Julius Caesar objection. In R. G. Heck (Ed.), *Language, Thought, and Logic. Essays in honour of Michael Dummett* (pp. 273–308). Oxford: Oxford University Press.

Hilbert, D., & Ackermann, W. (1928). *Grundzüge der Theoretischen Logik*. Berlin: Julius Springer. (Cited from 2nd ed. (1938))

Hilbert, D., & Bernays, P. (1934). *Grundlagen der Mathematik I* (No. 40). Berlin: Springer.

Kleene, S. C. (1952). *Introduction to Mathematics*. New York: Van Norstrand.

Klev, A. (forthcomming). The justification of identity elimination in Martin-Löf's type theory. *Topoi*. doi: 10.1007/s11245-017-9509-1

Locke, J. (1690). *An Essay concerning Human Understanding*. London: Thomas Basset. (Critical edition with annotations and introduction by Nidditch, P.H., published by Clarendon Press, Oxford, 1975)

Martin-Löf, P. (1975a). About models for intuitionistic type theories and the notion of definitional equality. In S. Kanger (Ed.), *Proceedings of the Third Scandinavian Logic Symposium* (pp. 81–109). Amsterdam: North-Holland.

Martin-Löf, P. (1975b). An intuitionistic theory of types: Predicative part. In H. E. Rose & J. C. Shepherdson (Eds.), *Logic Colloquium '73* (pp. 73–118). Amsterdam: North-Holland.

Martin-Löf, P. (1984). *Intuitionistic Type Theory*. Naples: Bibliopolis.

Morrill, G. V. (2011). *Categorial Grammar*. Oxford: Oxford University Press.

Sundholm, B. G. (1999). Identity: absolute, criterial, propositional. In T. Childers (Ed.), *Logica Yearbook 1998* (pp. 20–26). Prague: Filosofia.

Sundholm, B. G. (2002). What is an expression? In T. Childers & O. Majer (Eds.), *Logica Yearbook 2001* (pp. 181–194). Prague: Filosofia.

Wehmeier, K. (2017). Identity and quantification. *Philosophical Studies*, *174*, 759–770.

Wright, C. (1983). *Frege's Conception of Numbers as Objects*. Aberdeen: Aberdeen University Press.

Wright, C. (1998). On the harmless impredicativity of $N^=$ ('Hume's Principle'). In M. Schirn (Ed.), *Philosophy of Mathematics Today* (pp. 339–368). Oxford: Oxford University Press. (Cited from the reprint in Hale and Wright (2001))

Ansten Klev
Institute of Philosophy, Czech Academy of Sciences
The Czech Republic
E-mail: anstenklev@gmail.com

Proof in Mathematics and in Logic

DANIELLE MACBETH[1]

Abstract: What is the relationship between a mathematician's proof and the same proof as formalized in mathematical logic? The answer is not obvious insofar as something seems to be lost in the translation of the mathematician's proof into a fully formalized proof, which raises in turn the question how reasoning works in the actual practice of mathematicians. If we are concerned solely with valid rules of inference then the answer must be: by the same logic mathematical logicians use. The rules of valid inference are not in question. What is in question, I suggest, is how the system of written signs functions, how it is to be read.

Keywords: Formal proof, Machine reasoning, Mathematical logic, Mathematical notation, Mathematical reasoning, Mechanical reasoning, Diagrammatic reasoning

1 Introduction

Mathematical logic has bequeathed to us an image of mathematical reasoning according to which it is, in essence, nothing more than the mechanical manipulation of signs according to rules of logic—something that a machine might do as well as any human being. A properly rigorous, formalized proof

---

[1] Some years ago, Ken Manders asked me where the content goes when a mathematician's proof is formalized in mathematical logic, sparking the reflections that led ultimately to this essay. I have since given many talks on the subject: in 2012 at the University of Notre Dame, the inaugural meeting of the Philosophy of Mathematics Association; in 2013 at Indiana University, the annual meeting of the Society for the Study of the History of Analytic Philosophy; again in 2013 at the University of Illinois, Urbana-Champagne, the second international meeting of the Association for the Philosophy of Mathematical Practice; in 2014 at the Vrije Universiteit Brussel, the International Workshop on Logic and the Philosophy of Mathematics; in 2015 at the University of Kent, at the Society for the Study of Artificial Intelligence and the Simulation of Behavior; and most recently at Logica 2017, The Czech Republic. A much earlier version of this material, under the title Reasoning in Mathematics and Machines: The Place of Mathematical Logic in Mathematical Understanding, was posted online in the Proceedings of the 2015 meeting of the Society for the Study of Artificial Intelligence and the Simulation of Behavior. These opportunities to try out my developing ideas have been invaluable. I am especially appreciative of comments and questions from my audiences; these have been instrumental in my coming to more adequate formulations.

on this conception is one that is wholly gap free, a set of sentences some of which are premises and the rest of which can be derived from those premises or already derived sentences through the application of antecedently stated rules of logic governing the manipulation of signs in the system. There is, however, a problem. Not only do mathematician's proofs not conform to this image, proofs that do so conform are of little mathematical interest. They are not more rigorous than the proofs that mathematicians produce— at least by mathematician's standards of rigor. They do not provide insight and understanding. Often they are simply unintelligible. Mathematical logic seems, then, to be irrelevant to the reasoning of practicing mathematicians.

   Mathematicians in their practice have almost no interest either in mathematical logic or in proofs that conform to its standards of rigor. Does it follow that the claim that the mathematician's reasoning is just logical reasoning is false? This was the view of Poincaré when he was first introduced to Russell's formal logic by Couturat's writings on the subject in (Poincaré, 1905). Poincaré did not deny that there is a sense in which a mathematician's proof can be formalized. His claim was rather that the formalization lacks the *mathematics* of the original proof, that in formalizing one destroys the mathematical content, and thereby the mathematical interest, of the proof. This claim has also been defended more recently by Rav (1999). Others argue that we must understand mathematical reasoning as logical reasoning, as the mechanical manipulation of signs according to rules of logic. The features of proofs that mathematicians so esteem, such as the ability of some proofs to provide mathematicians with better understanding, are not, it is argued, matters of logic but belong instead to some other domain of inquiry. Goldfarb (1988), for example, suggests that issues of mathematical understanding and insight are not the concern of logic but belong instead to the psychology of mathematical thinking. Avigad (2006) suggests in a similar vein that what is needed to understand mathematicians' reasoning more fully is not logic but a philosophical investigation conducted alongside logic that is focused on the analytical study of mathematical methods of proof. Who is right? Is the mathematician's reasoning essentially different from the mechanical manipulation of signs as Poincaré thought? Or are mathematical and machine reasoning essentially the same as Goldfarb and Avigad hold? Given that the concern of logic is the goodness of inference, of deductive reasoning, and that mathematics, at least as it has come to be practiced since the nineteenth century, is primarily a matter of deductive reasoning from the contents of concepts, it can seem overwhelmingly obvious that it is Goldfarb and Avigad who are right: the image of reasoning that logic provides is

Proof in Mathematics

that of mathematical reasoning. And not only that: as Burgess (1992, p. 9) notes, "classical logic was developed ... as an extension of traditional logic mainly, if not solely, about proof procedures in mathematics." How, then, is it so much as *possible* that mathematical logic is, as it seems to be, irrelevant to reasoning in mathematical practice?

We know that the mathematician's proofs can be formalized, which suggests that mathematical reasoning is in essence nothing more than the mechanical manipulation of signs according to rules—again, something that a machine might do as well as a human being. But we also know that such formalizations are of no mathematical interest, which suggests, contrariwise, that mathematical reasoning is something essentially different from the mechanical manipulation of signs according to rules. Does the mechanical manipulation of signs perhaps only *mimic* the reasoning of the mathematician? I aim to show exactly that. More specifically, I will argue that although on one way of reading, the signs that are used in such cases of reasoning in mathematics appear to be manipulated merely mechanically, that is, as they are in mathematical logic, there is *another* way of reading those same signs that reveals the reasoning to be quite different from any mechanical manipulation of the signs. Specially devised mathematical systems of signs can be read merely mechanically, as they are in mathematical logic, but, we will see, they can also be read differently, as exhibiting the contents of mathematical concepts on the basis of which to reason. A mathematical language so read functions in an essentially different way from the way it functions in mathematical logic.

2   Two ways of reading

Imagine a person who knows how to count things but not any mathematical facts, even the most basic. Imagine further that this person wishes to know what is (say) the sum of seven and five. A good mechanical expedient for finding the answer would be, first, to count out seven things, then, to count out five more things, and finally, to count the resulting collection, something we can do with, for example, marks on a page: /////// /////. The collection on the left is a collection of seven strokes and that on the right is a collection of five strokes. By adding five more strokes after making the first seven strokes, I made, whether I knew it or not, a collection of twelve things—as I can learn by counting the whole. That there are twelve strokes is *already there*, contained implicitly in the display, as soon as five more strokes are added to the original seven strokes. What my subsequent counting does is only to make that fact explicit.

Interestingly, Kant seems not to think of the problem of the sum of seven and five in this way. (Kant discusses this case in the B Introduction of the first *Critique*, B15–16.) According to Kant, the truth that seven plus five is twelve is not analytic but instead synthetic a priori. But if so then being twelve is *not* contained already in the starting point, in the given numbers seven and five. Rather, Kant thinks, the number twelve can be *constructed* given the starting point. And we can see how this is to work if we conceive the strokes on the page not as mere *things*—marks on a page that perhaps also stand in for other things—but instead as primitive signs of an exceedingly simple mathematical language, as signs that express *Fregean senses*. We then can use these primitive signs to construct complex signs that designate numbers. On this *mathematical* reading, the collection of seven strokes is not merely a collection of marks on a page; it is instead a *complex sign* of a very simple mathematical language, a sign that although complex nonetheless designates *one* thing in particular, namely, *the* number seven, a certain mathematical object, and it does so through a Fregean sense that reveals that number as a certain multiplicity. Similarly, the collection of five strokes read as Frege would have us read is a complex sign designating *the* number five, again through a Fregean sense. On this reading, then, the individual strokes in the complex sign do not designate; only the whole complex sign, the whole collection of strokes read together as one sign, designates. And once we have two such signs, the complex sign for the number seven and the complex sign for the number five, we can construct a sign for the number twelve by combining all the primitive signs into one. This is a mathematical operation, not a merely mechanical one, insofar as, first, we are using the strokes to display the arithmetical contents of the numbers seven and five, what it is to be such numbers (namely, certain multiplicities); and subsequently, on the basis of that displayed content, we manipulate the primitive signs in a rigorous, rule-governed way in order to show something about the numbers involved, that the sum of the numbers seven and five is the number twelve, a properly mathematical result.

A different example again illustrates the essential point.

Suppose that you needed, for whatever reason, to construct a particular equilateral triangle, a material instance of that geometrical shape. You might proceed as follows. First, using a straight-edge, you draw a line the length of a side of the desired triangle. Now, using a compass, you draw two circles, one with one end of the drawn line as center and the line as

Proof in Mathematics

radius and the other with the other end of the drawn line as center and the line as radius. Then from one of the two points of intersection of the two circles, you draw, again using a straight-edge, two lines, one to each end of the original line. The three straight lines you have drawn form, more or less, a physical instance of a triangle all three sides of which are (more or less) equal in length, so the triangle you have drawn is an equilateral triangle (more or less). It is furthermore clear—given that your aim was to produce an actual instance of an equilateral triangle—that you were very well advised to use a straight-edge and compass so that your straight lines would be as straight as possible and your circles as circular as possible. Only by some such mechanical means will the sides of the particular triangle one draws approximate as closely as possible lines that are straight and all equal in length as required by the task that was set. One could of course also build a machine to perform such a task.

Consider now a *mathematical* demonstration of the construction of an equilateral triangle on a given straight line such as that Euclid provides in Proposition I.1 of the *Elements*. This demonstration uses essentially the same collection of marks on a page—with one crucial difference. One does not in the mathematical case need either a straight-edge or a compass, and one does not because one is not aiming to produce any particular triangle. The task is to demonstrate something *mathematical*, that it is possible, given a finite straight line, to construct an equilateral triangle on that line. The demonstration is general throughout. Of course, one does draw a particular (sort of straight) line, then two particular circles (or at least curves that look sort of circular), and then two more (sort of straight) lines. But it is not their actual shapes that matter to the mathematical reasoning. What matters is the way the drawn lines are intended to formulate conceptual content usable in the process of mathematical reasoning. In the context of reasoning in Euclid, a circle, for example, serves to license certain inferences: if one has two radii of one and the same circle then one can infer that they are equal in length—whether or not they *look* equal in length in the diagram as drawn! The proof is not mechanical, as it was in our first case involving a straight-edge and compass, but is instead mathematical: we establish that the sides of the triangle are one and all exactly equal in length not by measuring, which could never establish what is wanted, but by reasoning, in particular, by inferring that they are, again, exactly equal in length because they are all radii of mathematical circles of precisely the same size.

In the case of the construction of some one actual equilateral triangle, one draws lines in a step-wise fashion that eventuates in some physical thing that has a certain shape, more or less. The construction goes through, the triangle is produced, whether or not that which is producing it, whether human or machine, has any understanding of what it is doing. In the mathematical proof, although the array of lines may be drawn in a way that is visually indistinguishable from the array drawn in the case of the production of some actual triangle, there is no *proof* without the reasoning one does in the diagram. The mathematical proof lies in, first, judging that all three lines are and must be equal in length because they all are radii of two circles of precisely the same size. And having realized in this way that the three lines are equal in length, one must then see that the three lines, differently regarded, form a triangle, that what had been seen as radii of circles can *also* be seen as sides of a triangle. This perceptual skill of seeing something now one way and now another is essential to one's getting the proof as a proof of a mathematical result. In the mechanical construction the drawn lines are just that, drawn lines. In the mathematical proof, the lines serve a much more interesting and conceptually subtle function. In the context of Euclidean diagrammatic reasoning, a drawn straight line expresses a Fregean sense independent of any context of use; only given its place in a certain diagram and a particular way of regarding it, does it designate, say, a radius of a circle, or a side of a triangle.

Here, then, is the key idea. In a bit of mechanical reasoning, the signs, the written marks, are nothing more than counters to be manipulated in certain ways according to rules. They designate what they designate independent of any context of use or way of regarding them, and there is no need at all to take into account what it is that they designate in making the needed construction. A straight line is a straight line whether or not anyone notices, and a circle is a circle whether or not it is taken for one by anyone. In effect, we distinguish in this case between syntax and semantics, and need rely only on the former in our reasoning. Similarly, for the collections of strokes: there are as many as there are in a collection of them independently of anyone's knowledge of that fact. The fact that they may stand in for other things is, again, strictly irrelevant. In the case of mathematical reasoning the signs function very differently. They are not, in that case, merely counters to be manipulated according to rules. Instead the signs express Fregean senses. And independent of any context of use that is all that they do. Such signs, more exactly, collections of such signs, designate this or that mathematical concept only within a context of use and relative to a way of regarding them.

3 Reading logical notation

A collection of marks in our little stroke language or in a Euclidean diagram can be read either mechanically or mathematically. And the point generalizes: any system of written marks devised to display the contents of mathematical concepts in a way enabling rigorous reasoning in the system of signs can also be read merely mechanically precisely because the reasoning is rigorous, every step governed by an antecedently specified rule. Given a system of signs within which to display the contents of concepts in a way enabling rule-governed reasoning, one can invariably treat the primitive signs as mere marks, things to be manipulated as the rules allow without regard for what, if anything, they signify. And because one can, one can, in principle, build a machine so to manipulate the marks. What we need now to see is that just this same point applies also to the case of reasoning from the contents of *inferentially* articulated mathematical concepts. Although one *can* read the signs mechanically, as recording or picturing information (as we do in standard mathematical logic), one can also read them as Frege teaches us to read, as expressing Fregean senses that contain modes of presentation of the designated concepts. And having learned so to read the signs of logic, we will then be able to see that even strictly deductive reasoning, conceived as Frege conceives it, can be ampliative, a real extension of our knowledge. It can be ampliative because there is in such cases a kind of construction of the conclusion on the basis of the premises. By deductive reasoning on the basis of the contents of the concepts with which one begins, as given in their definitions, one proves theorems about those very concepts, about their logical relations one to another. The reasoning is not mechanical. It is mathematical reasoning, reasoning of the sort mathematicians actually engage in.

Language as it is understood in mathematical logic is a means of picturing facts; sentences of the language serve to record states of affairs, that is, truth-conditions, what is the case if the sentence is true. It follows that logical inference is to be understood on the model of containment: if the information in the conclusion is contained already in the premises taken as a whole, then the inference is valid. Venn diagrams provide a nice illustration of this containment conception of inference. First one pictures the information in the premises, say, that all A is B (by shading the area inside the A-circle but outside the B-circle to show that nothing is A but not B) and that all B is C (by shading the area inside the B-circle but outside the C-circle to show that nothing is B but not C). Having done that one can see in the diagram that one has thereby pictured also that all A is C. The area that is inside the A-circle but outside the C-circle has been entirely shaded in the course of picturing what is the case according to the premises. Actually to draw the inference, infer that all A is C, is merely to make that fact explicit.

Notice, now, the very strong parallel between what is going on in a Venn diagram and the case of determining mechanically what the sum of seven and five is using written marks as we first did. In both cases one displays or pictures something, namely, a fact or a collection of things, and one does so by generating a fact or collection. The collection of strokes is to be seen as just that, a collection of things, and, as is made explicit in Wittgenstein's *Tractatus*, a proposition picturing a fact or state of affairs, as on the reading given by standard logic, does so by itself being a state of affairs. (See, for example, §§ 3.1432 and 4.0311.) Because propositions serve on this account to picture facts by themselves being facts, just as a collection of strokes serves to picture a collection of things by itself being a collection of things, valid inference must be understood to be nothing more than a matter of making explicit information that was there implicitly already in the starting point. It is just as Wittgenstein says in the *Tractatus* § 6.1262: "Proof in logic is merely a mechanical expedient to facilitate the recognition of tautologies in complicated cases."

Frege's logical language *Begriffsschrift*, first introduced in Frege's 1879 monograph *Begriffsschrift*, functions very differently. As I show in my 2005 book *Frege's Logic*, *Begriffsschrift* is not designed to record or picture truth-conditions; it is designed to enable the expression of a content, specifically the contents of concepts in definitions on the basis of which to reason. In my 2014 book *Realizing Reason* I further show how Frege's proof of theorem 133 on the basis of four definitions in Part III of the 1879 logic reveals the extraordinary power of the language. This proof is ampliative despite being strictly deductive—something that, as Frege explicitly notes in *Grundlagen* § 88, is obviously incoherent on a Kantian understanding of deduction. Whereas deduction as Kant understands it is analytic, by contrast with the sort of constructive and ampliative reasoning one finds in mathematics on Kant's view, deduction in Frege can be ampliative, that is, synthetic rather than analytic, and hence in its way constructive just as Kant thinks mathematical reasoning is. To understand how this is to work, at least in outline, we need to think about the nature of definitions in mathematics.

According to the Kantian, according, that is, to anyone for whom standard mathematical logic just *is* logic, a definition serves to introduce an abbreviation: on the one side, in the definiens, one sets out necessary and sufficient conditions for the application of some notion, what is true if and

Proof in Mathematics

only if the notion obtains; and on the other side, in the definiendum, one introduces a new simple sign to abbreviate those conditions. If, then, one proves something on the basis of some definitions through the rule-governed manipulation of signs, that can only be because the information in the conclusion is implicitly contained already in those definitions taken together.

Frege's conception of definition is very different. According to Frege, the definiens serves to display a Fregean sense, the inferentially articulated content of some concept that is thereby designated, the same concept that is designated also by the simple sign, the definiendum, though through a very different sense. What we have in a Fregean definition, then, is something essentially like our collection of strokes on the *mathematical* reading. We have in the definiens a complex sign designating some concept, a complex sign that, because it is complex, can be manipulated according to rules just as the complex signs for the numbers seven and five could in our earlier example. And in the case of definitions, such manipulations can reveal logical bonds among the concepts defined. In essence, just as the complex signs for the numbers seven and five provide, on the mathematical reading, everything one needs in order to construct a complex sign for the number twelve, so the definitions of mathematical concepts, read as Frege would have us read, can provide everything one needs in order to construct a complex sign showing their logical relations one to the other. Here one does not merely make explicit what was implicit already in one's starting point. Instead one realizes or actualizes what is otherwise merely potential. In Frege's imagery in *Grundlagen* § 88, the conclusion is "contained in the definitions, but as plants are contained in their seeds, not as beams are contained in a house". Beams are implicitly contained in a house; they are already there and can be taken out, made explicit, by dismantling the house. A plant is not contained in the seed in this way. A plant can be grown from a seed, but it is not already there in the seed. Rather, the plant actualizes a *potential* that is in the seed. And so it is in an ampliative deductive proof according to Frege: what the reasoning does is to actualize a potential that is contained in the definitions with which one begins. The reasoning is ampliative despite being strictly deductive.

4 Conclusion

According to Poincaré, the formalization of a mathematician's proof in standard logic destroys the mathematics of the proof. This can seem very puzzling: where does the mathematical content *go*? The problem, we have seen, is not merely that one formalizes in the sense of providing fully gap-free proofs by appeal only to antecedently specified rules. For, as we know, we have developed systems of written signs within which to exhibit the contents of mathematical ideas in ways that are mathematically tractable, in ways that enable what even mathematicians agree is fully rigorous, rule-governed reasoning in the system of signs. The problem with formalizations in standard logic is that in that case one *also* mechanizes. One reduces the mathematician's reasoning to the merely mechanical manipulation of signs, and this does indeed destroy all that was mathematically significant about the proof. We begin, that is, with a mathematical language such as, for instance, the system of diagrams in Euclid or Frege's logical language *Begriffsschrift* within which to reason deductively on the basis of the contents of concepts. But instead of reading the language mathematically, we (mistakenly) read it mechanically; and because we do, we think that the translation, say, of Frege's *Begriffsschrift* notation into standard logical notation is nothing more than a notational convenience. But it is not. The original mathematical language, although it can be read mechanically, was to be understood as properly mathematical, not merely as a means of recording information. And once the mechanical reading has been given of the original language, and a translation then made into standard notation, the original mathematical content is unrecoverable, just as Poincaré saw.

Mathematicians often reason in specially devised written systems of signs, for example, in Euclidean diagrams or in the symbolic language of arithmetic and algebra. Such systems of signs furthermore function in the mathematician's use of them as Fregean languages enabling the display of conceptual content. Because the primitive signs of such languages only express Fregean senses independent of any context of use, complex signs of these languages can serve to exhibit the contents of mathematical concepts in mathematically tractable ways, in ways enabling rigorous, rule-governed reasoning *in* the system of signs. But what we have also seen is that such systems of signs can invariably be read also mechanically, each primitive sign as designating or standing in for what it means independent of any context of use. And this, I have indicated, is precisely what happened in the case of Frege's logical language *Begriffsschrift*. Although Frege designed it to be read mathematically, it was in fact read, first by Russell and then by all the rest of us following him, merely mechanically. Thus, it has come to seem that the mathematician's reasoning is really merely mechanical, something a computer can be programmed to do as well as any human being. But that is a mistake. Although the machine can mimic reasoning, it is not actually reasoning, not in anything like the sense in which we reason. The mechanical manipulation of signs according to rules in a proof that has been formalized in mathematical logic really is essentially different from reasoning on the basis of an understanding of concepts.

References

Avigad, J. (2006). Mathematical method and proof. *Synthese, 153 (1)*, 105–159.

Burgess, J. (1992). Proofs about proofs: A defense of classical logic. Part I: The aims of classical logic. In M. Detlefsen (Ed.), *Proof, Logic and Formalization* (pp. 8–23). Routledge.

Euclid. (1956). *The Thirteen Books of Euclid's Elements*. New York: Dover.

Frege, G. (1972). Conceptual notation: A formula language of pure thought modeled upon the formula language of arithmetic [*Begriffsschrift*]. In T. W. Bynum (Ed.), *Conceptual Notation and Related Articles* (pp. 101–203). Clarendon Press.

Frege, G. (1980). *Foundations of Arithmetic [*Groundlagen*]*. Evanston, Ill.: Northwestern University Press.

Goldfarb, W. (1988). Poincaré against the logicists. In W. Aspray & P. Kitcher (Eds.), *Minnesota Studies in the Philosophy of Science* (Vol. XI, pp. 61–81). University of Minnesota Press.

Kant, I. (1998). *Critique of Pure Reason*. Cambridge: Cambridge University Press.

Macbeth, D. (2005). *Frege's Logic*. Cambridge, Mass.: Harvard University Press.

Macbeth, D. (2014). *Realizing Reason: A Narrative of Truth and Knowing*. Oxford: Oxford University Press.

Poincaré, H. (1905). Les mathématiques et la logique. *Revue de Métaphysique et de Morale, 13*, 815–835.

Rav, Y. (1999). Why do we prove theorems? *Philosophia Mathematica, (III) 7*, 5–41.

Wittgenstein, L. (1961). *Tractatus Logico-Philosophicus*. Oxford: Basil Blackwell.

Danielle Macbeth

Danielle Macbeth
Haverford College
Department of Philosophy
Pennsylvania, USA
E-mail: dmacbeth@haverford.edu

Laws of Logic – Where Do They All Come From?

JAROSLAV PEREGRIN AND VLADIMÍR SVOBODA[1]

Abstract: Everybody would probably agree that there are various laws of logic, such as the law of (non-)contradiction, the law of the excluded middle, *modus ponens*, *ex falso quodlibet* and so on. It is however unlikely that everybody would agree on which of these laws are the *genuine* laws, in that they are nonnegotiable. But first and foremost there is almost no agreement with respect to *the nature* of the laws, what exactly the laws are about, which domain they regulate and what is the source of their authority. This is quite surprising, for the laws appear to lie within the very foundations of logic. In this paper, we summarize a very down-to-earth and naturalistic explanation of the nature of logical laws that stems from the account of their constitution that we presented in our recent book.

Keywords: Logic, Logical laws, Logical forms, Reflective equilibrium

1 Arguments and their correctness

The term "law" is ambiguous. We surely mean something different when we speak about *laws* in legal discourse, when we speak about the *laws* of physics and when we speak about *logical laws*. And while talking about laws in jurisprudence and in the natural sciences is quite common, talking about the laws of logic within modern philosophy and specifically within modern logic may sound somewhat obsolete. Philosophers used to conceive of the laws of logic as the cornerstones of (rational) thought, but when logic, thanks to Frege and others, became independent of psychology the laws ceased to be associated with processes that occur (or ideally should occur) in the minds of thinkers. In modern logic, there is no single law that has remained unchallenged;[2] logicians may not even use the term "logical

---

[1] Work on this paper has been supported by Research Grant No. 17-15645S of the Czech Science Foundation.

[2] The rejection of the most solid cornerstone of traditional logic – the law of (non-)contradiction – is constitutive of the currently flourishing enterprise of paraconsistent logic.

laws" anymore (preferring terms like "principles", "rules", etc.), and they are also less prone to talk about *thinking* but more prone to talk about *reasoning* (which is not necessarily understood as a mental process), *arguments* or *proofs*. But independently of how we choose to call the principles that logic is after, the question of the nature and origins of these items seems important. In this paper, we will try to provide an answer.

Let us start our inquiry with some preliminary observations that shouldn't be controversial:
1. *Logic's primary business is with arguments/reasoning.* This is not to say that logicians aren't studying other issues, indeed they are; but studying arguments is the most substantial *raison d'être* of logic, and the other issues logicians deal with unfold from this.
2. *Arguments consist of meaningful sentences (or perhaps meanings of the sentences).* Steps like

(A1) *If it rains, the streets are wet*
      It rains
      The streets are wet

or

(A2) $(1+1=2) \vee (1+1=3)$
      $\neg(1+1=3)$
      $1+1=2$

are arguments. Of course we can have "arguments" like

(AF1) $A \rightarrow B$
      A
      B,

but insofar as A and B are not mere shortcuts for particular meaningful sentences, this is just an argument *scheme*.
3. *Logic is concerned with the correctness of arguments/reasoning*[3]. Indeed, logic is concerned with telling us that (A1) is a correct argument, while

---

[3] In this article, similarly as in Peregrin and Svoboda (2017), we are going to follow a terminological convention – we will speak about *correctness* and *incorrectness* in the case of (full-fledged) arguments and about *validity* and *invalidity* in the case of argument schemes (forms).

(A3) *If it rains, the streets are wet*
      The streets are wet
      It rains

is not.[4]

4. *Especially, logic is supposed to identify arguments which are correct merely due to their "logical form".*[5] (Though there is no general agreement w.r.t. what a logical form is.)

These points all sound quite plausible; nevertheless, they constitute the point of departure for the—*prima facie* perhaps not so plausible—story about the nature of logical laws that we are going to tell. Let us add one more point which might seem slightly more controversial, namely:

5. *Not all correct arguments are logically correct.* What we mean by this is that besides logically correct arguments, like (A1) or (A2), we can also have arguments that are analytically correct, such as, for example

(A4) *Tom is a bachelor*
      Tom is male

or

(A5) *It is Monday today*
      It will be Tuesday tomorrow

and even arguments that are correct in a somewhat less definite sense, like

(A6) *Tom is in Slovakia*
      Tom is in Europe

or

(A7) *Bolzano was born in 1781*
      Bolzano was born earlier than Frege

We suggest calling arguments like (A6) and (A7), which are commonly taken as correct but in whose case it is perhaps thinkable that they could

---

[4]It is worth noting that showing which arguments are logically *incorrect* is a process incomparably more tortuous than showing which ones are *correct*; see Svoboda and Peregrin (2016).

[5]This task is sometimes alternatively formulated as the task of identifying logical truths but this alternative formulation is, in our view, potentially misleading.

be incorrect (providing the world were quite different from what it is), *status quo* correct.⁶ It is quite clear that logic is not supposed to demonstrate the correctness of arguments which are (only) analytically or *status quo* correct.⁷ Let us now consider some more examples of arguments with which we can be confronted in real life communication:

(A8) <u>All transcendental numbers smaller than 1 are irrational</u>
Some transcendental numbers smaller than 1 are irrational

(A9) <u>Tom knows that gold is necessarily heavier than aluminum</u>
Aluminum can't be heavier than gold

(A10) <u>Amundsen flew to the North Pole in his airplane</u>
Amundsen flew somewhere

(A11) <u>Tom ought to learn Russian</u>
Tom ought to learn German or Russian

Obviously, in the case of each of these arguments we can ask whether it is *logically correct*. Clearly answering this question can be a worthy task – we often need to decide whose argumentation is conclusive, or at the very least we need to secure a mutual understanding among communicating people. And we will want to know whether, in a particular case, we can do this with the help of logic. But who is supposed to be qualified to decide whether arguments like (A8) – (A11) are logically correct?

The natural assumption seems to be that it is the job of logicians. We, however, presume that this answer wouldn't be generally adopted. Somebody, having in mind what present day logicians actually do most of the time, might maintain that logicians are not really supposed to answer questions like this. They operate within the realm of the formal (they especially work with artificial languages that can be treated purely mathematically). How the forms relate to real languages and real arguments is perhaps a matter to be left to some of the more applied scientists (perhaps linguists or specialists in communication studies?). The trouble, as we see it, is that if logicians are supposed to dwell in the realm of the formal, there is nothing that distinguishes them from mathematicians; especially, there is nothing

---

⁶Of course, we do not claim that in natural languages the boundary lines between different kinds of correct arguments are sharp.

⁷For more about the classification of different kinds of correct arguments, see Peregrin and Svoboda (2017, chapter 2).

that makes them prone to address argumentation/reasoning as it actually takes place.

According to us, however, it is the predicament of logicians to deal with real arguments – for verdicts on them are *constitutive* of their business. Of course, logicians often use mathematical methods and they concentrate on the *logical forms* of arguments. But the ultimate topic of their study is the correctness of fully-fledged argumentation/reasoning.

2   Logical forms and logical laws

One of our initial observations was that logicians concentrate on logical forms. Prima facie this claim sounds perspicuous and uncontroversial, but it is fully intelligible only provided the concept of logical form is clear. Again, we believe that there are some quite uncontroversial points on which a vast majority of logicians would agree:

1. *Logical forms can be ascribed to meaningful sentences.* If logicians are to fulfill the task of deciding which arguments are correct they have to determine the logical forms of the involved sentences; for example, to ascribe a logical form to sentences like *The king of France is bald*, *Tom knows that gold is necessarily heavier than aluminum* or *Amundsen flew to the North Pole in his airplane.*

2. *Logical forms are articulated in various artificial languages employed by logicians.* To say what the logical form of a sentence is we employ a formal language. (Logicians usually present the logical form of *The king of France is bald* or *Amundsen flew to the North Pole in his airplane* in the language of predicate calculus, while that of *Tom knows that gold is necessarily heavier than aluminum* in the language of modal/epistemic propositional logic.)

However, when it comes to the *nature* of logical forms, controversies start.

A possible position is that logical forms are real – independent of any languages used by humans. They may be said to display the logical (or more broadly formal) structure of the world (which we are able to recognize/view/ recollect), or they may be seen as principles underlying thought as such. In both cases, logical languages are only our (imperfect) tools used to bring them to light (hence, there must be something that is *the* objective formal structure of the world or thought and consequently something like *the* logical language that captures the forms precisely).

We are skeptical about such a picture of logical forms. We, of course, don't think that such a position can be refuted by empirical research or decisive argumentation; but we are on the other hand also convinced that there is no evidence that would support such an "absolutist" stance. Thus, we suggest that we should base our discussion of logical forms on somewhat less arcane grounds – on the reflection of how logicians actually proceed when they are asked to identify the logical form of a sentence or an argument.

We are convinced that a kind of *instrumentalist* and *relativist* view is more plausible: the concept of a logical form of a natural language sentence makes, in our view, sense only relative to a logical language. The logical form of a sentence is the "best" way of capturing the inferential properties of a natural language sentence in a given logical language. It follows that *to have logical forms, we must have logical languages*. Now, as logical laws, as we have suggested, are also a matter of logical forms, fully-fledged and definite logical laws are unthinkable without artificial logical languages.[8]

An obvious objection is that logical laws should be inherent to argumentation/reasoning (even constitutive of it) rather than emerging only within the tools invented by logicians. Our reply is that relevant laws (which may be called proto-logical) do govern our practices of argumentation and reasoning (and thereby get sedimented within our natural languages), but in the form of merely *implicit* rules, which take the shape of definite principles only when they are fixed within a logical language. Only then do they become *logical laws* worth the name.

Now, it seems, we must pause and say something more about the nature of logical languages. We are convinced that artificial languages formed by logicians can be plausibly seen as specific *models* of de facto argumentation/reasoning. Similarly as mathematical models that physicists employ when they want to understand (e.g.) fluid mechanics, logical languages act as models that we use to get a better grasp of *de facto* argumentation/reasoning (and possibly to enhance it). In both cases, if we have "good" models (which proved useful/adequate enough) we can carry out lots of investigations inside the models.[9]

Artificial languages of logic are similar to scientific models of natural phenomena in that, in both cases, the ultimate end is to project the results achieved by the mathematical study of the models back on the original subject matter of the investigation – real cases of moving fluids or real argu-

---

[8] See Peregrin (2010b).
[9] Seeing the artificial languages of logic as such kinds of models is in no way unprecedented. It has already been suggested by Burgess (1992) or Shapiro (2001). The authors, however, see them primarily as models of mathematical practices, whereas we are convinced that if logic is to live up to its task, it should reflect our argumentative practices from a more general perspective.

mentation/reasoning. But, unsurprisingly, there is a vital difference. While in case of physics this projection in no way changes the original subject matter, in case of logic it may – there may be a feedback from logical theories on our practices of reasoning. Logical models are not purely descriptive or predictive, they are expected to *fix the rules* governing real argumentation.

It follows that the laws of logic(s) are not discovered – they are more in the nature of artifacts than excavations. Of course they are not created arbitrarily, they are rooted in natural languages, which are the fount of all meaning (worth the name). Laws like *modus ponens*, *excluded middle* and possibly more complex and less pronounced principles governing the 'serious discourse' (within which people are aiming at a consensus based on mutual understanding and at the extending of their knowledge) emerge as a *reflective equilibrium* (Brun, 2014; Peregrin & Svoboda, 2017). They result from our back and forth movement between the "data" (facts of argumentation/reasoning) and a "theory" (tentative articulations of rules constitutive of an artificial language that aims to attain the status of a logical language).

It is worth mentioning that the term "logical language" is somewhat ambiguous. One possibility is to view a logical language as defined by its formation rules. (Then we will say that, for example, classical propositional logic and intuitionistic propositional logic, or modal logic S2 and S5, share the same language.)[10] But we can also view logical language as not only defined by formation rules but also by transformation rules – by the axioms which establish inferential relations among its formulas/sentences or by means of a formal semantics. In this second and third sense, delineation of a logical language delineates a logic (the terms "logical language" and "logical system" or "logical calculus" are synonymous). This is also the sense in which we use the term here.

But there is another ambiguity that enters the picture when we use the term "logical" in this way. We can conceive of logical languages as being purely formal, their extralogical terms being contentless parameters and the formulas containing them thus being unable to express sentences with full-fledged meaning, or we can conceive of them as "fully-interpreted" languages whose extra-logical terms are meaningful constants whose formulas are (or at least can be) meaningful sentences.[11] This ambiguity is not so important for us here.

---

[10]We should note that even this "syntactical" concept of logical language is not purely formal. We naturally assume that in two logical languages which generate the same formulas the same symbols (logical constants) are associated with the same logical expressions of natural languages – for example, English expressions like "and", "every", "not" or "possibly". If they are not we wouldn't speak about the same language.

[11]For more about this, see Peregrin and Svoboda (2017, chapter 4).

Jaroslav Peregrin and Vladimír Svoboda

3 Reflective equilibrium

Let us consider, once more, the argument

(A1) *If it rains, the streets are wet*
      It rains
      The streets are wet

We readily classify it as correct. Why? In our view it is because we were taught to use expressions which form it, in particular "if", in a certain way. (In fact, we would see assenting to inferences of the kind of (A1) as *touchstones* of understanding "if".) We are also sure that it would be accepted as correct by (the great majority of) the speakers of English.[12] Also, we don't doubt that the speakers would also accept as correct similar arguments like

(A12) *If you don't have any money, you cannot buy the cake*
       You don't have any money
       You cannot buy the cake.

Why are these arguments similar? The answer is not difficult – they are similar because they have the same form, namely

If A, B
A
B.

Now if we introduce a convention and decide to write "⇒" instead of "if" we get

⇒ *A, B*
A
B.

This form of (English) arguments is valid. Why? Because, we assume, all of its instances are arguments which would be considered as correct by English speakers who understand "if" and read "⇒" as its shortcut. (Well, in fact *almost* all, for some of the instances may sound weird, and there may

---

[12] This is not to say that for *any* arguments the acceptance by the majority would be equal to its correctness. However, this holds for simple and perspicuous arguments of this kind.

perhaps even be instances which would be classified as incorrect by a substantial portion of the speakers.[13]) If we now introduce one more convention and decide to write "$\Rightarrow$" in between two clauses instead of in front of them, we have the familiar scheme[14]

(MP) $\quad \dfrac{\begin{array}{l} A \Rightarrow B \\ A \end{array}}{B.}$

Let us now consider a similar form articulated in the language of a logical system S:

(MP*) $\quad \dfrac{\begin{array}{l} A \rightarrow B \\ A \end{array}}{B}$

The question whether it is valid cannot of course be answered unless we are acquainted with the system S. But, influenced by our expectations, we will tend to assume that it is *valid* (we would assume that "$\rightarrow$" is an implication sign and we know that this scheme is valid in classical logic, as well as in many other logical calculi). What is important to keep in mind is that if the scheme of (the language of) S is valid, it is because of the definition of "$\rightarrow$" in S – its validity is in this sense a (trivial) consequence of certain given postulates.

Now, can either (MP) or (MP*) be identified with the law of *modus ponens*? Let us first consider the second possibility. In such a case we would, as it seems, have to admit that *modus ponens* is just a trivial consequence of mathematical definitions which govern, within S, the use of "$\rightarrow$". If, on the other hand, *modus ponens* is to be identified with (MP), then the validity of this law would depend on empirical facts – facts that have been brought about by the contingent development of English (plus our contingent convention concerning the use of "$\Rightarrow$") and that might come to be contravened by its further development.[15] Neither possibility looks quite plausible.

Doesn't this (admittedly sketchy) reflection on the two options suggest that it might be, after all, most reasonable to retreat to the view which places *modus ponens* somewhere beyond any (natural or artificial) languages, in

---

[13] In the literature, we can come across arguments which seem to shake overly bold claims about the validity of all arguments of this form (see, e.g., McGee 1985).

[14] The previous scheme would perhaps look more familiar to some Polish logicians.

[15] Cf. Peregrin (2010a).

some realm of forms of the world or of thought? Aren't principles like *modus ponens* something that resides in the ideal world of the purely formal? Maybe we, humans, were designed as rational beings or we have developed into beings with principles like *modus ponens* imprinted into our minds (and so we can recognize them *a priori*).

Though contemplations of this kind may sound tempting to many ears, we don't think that they are promising if we are searching for an understanding of the foundations of logic. The "solution" they offer is, in our view, illusory – not only because it relies on tricky metaphysical assumptions (which are sometimes hardly distinguishable from a wishful thinking), but especially because the assumption that the genuine implication is an abstract, ideal object does not solve anything. The Platonist heaven, if we admit its existence, abounds in all kinds of objects (they certainly contain classical implication, intuitionist implication, various relevance implications and many others which are not incorporated in any logical system that we have created so far), and it is quite unclear how to answer the question of which one is the mythical *genuine implication* (unless we want to say that it is the one that is expressed by a natural or a distinguished artificial language, in which case we are back at one of the kinds of answers we rejected above).

What we suggest is that the "genuine implication" is not a kind of esoteric object beyond any languages, but rather a result of a complex interplay between elements of natural and formal languages. Consider how logical systems – like our generic system S – get formed. Certainly not every language and every calculus we can put together deserves to be called *logical*. Systems that deserve such a designation emerge from a process of the complex, mutual adaptation of an artificial language (not necessarily the rich artificial language like those we know from modern logic, but possibly also a proto-artificial language, such as that of Aristotelian syllogistics) and a natural one; they result from our pre-theoretical language meeting with its systematic theoretical reflection. As we have suggested (Peregrin & Svoboda, 2017), the process of formation of such logical languages/systems (which in practice can take different shapes) has the character of a *reflective equilibrium*.[16]

---

[16]The term "reflective equilibrium", which is now commonly used as a name of the method/process as well as the name of its outcome, became widely used after the publication of Rawls' influential book *Theory of Justice* (Rawls, 1971). The idea (conceived more broadly than in the case of Rawls, who focuses on ethics) goes back (at least) to Goodman (1955). A formal analysis of the process is presented by Brun (to appear).

Laws of Logic

The idea is that the development of our languages was intertwined with the development of our argumentation/reasoning. Similarly as in ontogenesis, the process of learning to reason is inseparable from the process of mastering a language; the two processes, we may plausibly assume, were also inseparably connected in human phylogenesis. Natural languages have developed as governed by certain implicit rules, rules that are fuzzy and open-ended. Our theoretical reflection on these rules (which is often motivated by the need to make the language less fuzzy and more exact) resulted in our positing their explicit, crisp and closed explications. These posits get confronted with their natural counterparts and get amended where this confrontation yields overly large discrepancies. However, it is not only the posited explicit rules but also the underlying implicit ones that may get amended by the confrontation; hence, what occurs is a back-and-forth movement between the tentative theoretical generalizations and our "intuitions" underlying them. And this movement proceeds towards the kind of equilibrium which yields a theoretical tool that will serve our purposes.

As there is no one "true" logical system, there is no one "true" implication and no one "true" *modus ponens*. We have, strictly speaking, distinctive versions of *modus ponens* as articulated in different languages which have qualified as *logical*[17] (and which, at the same time, use a junction designed for straightforward formalization of conditional sentences).[18] However, there is clearly a sense in which the different versions of implication can still be seen as different species of the same kind[19] and hence we have also a general notion of *modus ponens*: it is the rule which takes us from an indicative conditional plus its antecedent to its consequent.

What is remarkable is that if we have an artificial language which has stood the test of reflective equilibrium considerations (and hence deserves the name *logical*), we can make use of its constants in a specific way. We can formulate full-fledged arguments which combine such constants (whose meaning is precisely determined) with expressions of natural languages such as English. Thus, we can identify instances of logical laws which are in a way very special. They belong neither to a logical language nor to a nat-

---

[17] Logicality, within this picture, is not a "yes or no" matter, it may come in degrees. For example, systems with a very limited expressive power can still count as logical if they are useful for some purposes, though generally we expect that logicality presupposes a kind of versatility.

[18] We wouldn't call the argument form $\neg A \vee B$, A hence B a specific version of *modus ponens* though it is in classical logic indistinguishable from $A \rightarrow B$, A hence B.

[19] Perhaps the situation can be compared to law – though there is nothing like one, *single* law prohibiting rape, it sounds plausible to say that we can find, in different countries, specific versions of such a law.

ural one, but they are formulated in a *hybrid language* resulting from their crossover. Thus we can, for example, "crossbreed" the language of classical propositional logic with its connective "→" and English and formulate the following argument:

(AH3) *It rains → the streets are wet*
It rains
The streets are wet

This "hybrid" argument is clearly an instance of (a specific version of) *modus ponens*. It can be seen as both *full fledged* (meaningful, fully understandable) and *undeniably correct* (logicians who "talk the language" of classical propositional logic are the ultimate arbiters concerning the correctness).[20] In this (and not only in this) way logic(s) allow us to move the expressive potential of our language to a new level and assure (at least to some extent) firm common grounds for our discussions.

4 Conclusion

Laws of logic interconnect logical forms, while logical forms are our theoretical reconstructions of the inferential properties of the sentences we use to reason/argue, *viz.* typically declarative sentences of our natural language. While our practices of argumentation/reasoning are rule-governed in the sense that we do correct each other and thus reinforce what can be seen as "implicit rules" inherent to the enterprise, genuine, explicit rules originate only from our theoretical reflection of the practice. The rules are formed by a process in which our attempts at fixing the rules in an explicitly articulated form get accommodated to the "intuitions" which underlie our "implicit rules", while the "intuitions are amended by the emerging rules – the process of *reflective equilibrium*. Thus, logical laws are neither rules concerning directly our natural languages nor precepts embodied in definitions of a certain artificial language. But neither do they concern some *a priori* given forms beyond any language. They result from a delicate interplay between natural languages with their implicit rules and artificial languages with their stipulated explicit rules, which results in rules that are rooted in our natural languages (and the practices of argumentation/reasoning which they are the vehicle of) but which are, however, explicit and open to view.

---

[20] If we used (certain) strict implication or relevant implication in place of the material implication we would, of course, receive a somewhat different argument as the meaning of the first premise would be different.

References

Brun, G. (2014). Reconstructing arguments: Formalization and reflective equilibrium. *Logical Analysis and History of Philosophy, 17*, 94-129.

Brun, G. (to appear). Conceptual re-engineering: from explication to reflective equilibrium. *Synthese*.

Burgess, J. (1992). Proofs about proofs: A defense of classical logic. In M. Detlefsen (Ed.), *Proof, logic and formalization* (pp. 8–23). London: Routledge.

Goodman, N. (1955). *Fact, Fiction, and Forecast*. Cambridge (Mass.): Harvard University Press.

McGee, V. (1985). A counterexample to modus ponens. *Journal of Philosophy, 82*, 462-471.

Peregrin, J. (2010a). Logic and natural selection. *Logica Universalis, 4*, 207-223.

Peregrin, J. (2010b). The myth of semantic structure. In P. Stalmaszczyk (Ed.), *Philosophy of Language and Linguistics, vol. I: The Formal Turn* (pp. 183–197). Frankfurt: Ontos.

Peregrin, J., & Svoboda, V. (2017). *Reflective Equilibrium and the Principles of Logical Analysis: Understanding the Laws of Logic*. New York: Routledge.

Rawls, J. (1971). *A Theory of Justice*. Cambridge (Mass.): Harvard University Press.

Shapiro, S. (2001). Modeling and normativity, how much revisionism can we tolerate? *Agora, 20*, 159-173.

Svoboda, V., & Peregrin, J. (2016). Logically incorrect arguments. *Argumentation, 30*, 263-287.

Jaroslav Peregrin
Czech Academy of Sciences, Institute of Philosophy
The Czech Republic
E-mail: peregrin@flu.cas.cz

Vladimír Svoboda
Czech Academy of Sciences, Institute of Philosophy
The Czech Republic
E-mail: svobodav@flu.cas.cz

Ranking Semantics for Doxastic Necessities and Conditionals

ERIC RAIDL[1]

Abstract: Ranking semantics is the implementation, in a possible world setting, of Spohn's (1988, 2012) ranking theory. This article proves completeness results for the nested doxastic necessities, states interactions with the graded conditionals and discusses some possible applications and extensions.

Keywords: Ranking theory, Conditionals, Graded modality, Ranking semantics, Normal modal logic, Linearly ordered modal logic, Reason relation

Introduction

Ranking semantics for conditionals was one of the plausibility semantics introduced in Friedman and Halpern (2001), generalised as multi-agent semantics by Halpern (2003). Spohn (2015) develops a specific version for a non-nested fragment of the conditional language. Huber (2014, 2015, 2017) discusses philosophical motivations and justifications, and proposes modifications of the truth clause of the conditional. Completeness results for these modifications are proven in Raidl (ms), with a discussion of doxastic impossible antecedent conditionals. These works only investigate the behaviour of the 'null-conditional' in the ranking semantics. More recently, Lauer (2017) has used the ranking semantics to model graded belief as graded modality. The paper focuses on the latter, which are in fact the inner modalities of the graded conditionals. Completeness results are proven for these modalities.

§1 introduces ranking theory, §2 defines the ranking semantics, §3 provides possible applications to natural language as well as to epistemology and §4 contains the completeness results.

---
[1] I thank Wolfgang Spohn, Niels Skovgaard-Olsen, Sven Lauer, and Franz Huber for stimulating discussions, members of the 'what if?'-group and participants of Logica 2017 for their thoughtful remarks.

Eric Raidl

1 Some ranking theory

This section provides the basic definitions of ranking theory.[2]

Write $\mathbb{N}_\infty := \mathbb{N} \cup \{\infty\}$. k is a *(natural valued) ranking mass* iff it is a function $k : W \longrightarrow \mathbb{N}_\infty$, such that $k^{-1}[0] \neq \emptyset$. A *(complete) ranking function* over $\wp(W)$[3] is generated from k by setting $\kappa_k(\emptyset) := \infty$ and $\kappa_k(A) := \min_{w \in A} k(w)$ for $A \in \wp(W) \setminus \{\emptyset\}$.[4] k is *regular* if $k[W] \subseteq \mathbb{N}$ and κ_k is *regular* if k is. The *conditionalisation* of κ is defined for all $C \in \wp(W)$, s.t. $\kappa(C) < \infty$ by setting $\kappa_C(A) := \kappa(A \cap C) - \kappa(C)$ for all $A \in \wp(W)$. κ_C is again a ranking function, provided κ was and $\kappa(C) < \infty$. Conditionalisation is *vacuous* for W, i.e., $\kappa_W = \kappa$. For $x \in \mathbb{N}_\infty$ and κ a ranking function, x-*belief* and *conditional x-belief* are defined by: $\mathbf{B}^n_\kappa(A)$ iff $\kappa(\overline{A}) > n$, $\mathbf{B}^\infty_\kappa(A)$ iff $\kappa(\overline{A}) = \infty$, and $\mathbf{B}^x_\kappa(A|C)$ iff $\mathbf{B}^x_{\kappa_C}(A)$, provided $\kappa(C) < \infty$. The belief notion used in Spohn (2012, 2015) is the null-belief (0-belief) and the conditional belief underlying the ranking semantics used in Friedman and Halpern (2001), Halpern (2003), as well as in Huber (2014, 2015, 2017) and in Spohn (2015) is the conditional null-belief. Lauer (2017) uses strength indexed belief, but only needs to distinguish two values. Here, we consider the full range of graded belief and conditional belief.

For κ a complete ranking function and $x \in \mathbb{N}_\infty$, define $[\leq x]_\kappa := \{w \in W : \kappa(\{w\}) \leq x\}$ and $[< x]_\kappa := \{w \in W : \kappa(\{w\}) < x\}$. We have

Lemma 1 (belief) *Assume $\kappa(C) < \infty$.*

1. $\mathbf{B}^x_\kappa(.|W) = \mathbf{B}^x_\kappa$.

2. $\mathbf{B}^x_\kappa$ *(resp. $\mathbf{B}^x_{\kappa_C}$) generates a proper filter.*

3. $\mathbf{B}^n_\kappa(A)$ *iff* $[\leq n]_\kappa \subseteq A$. $\mathbf{B}^\infty_\kappa(A)$ *iff* $[< \infty]_\kappa \subseteq A$.

4. *For $y > x$, $\mathbf{B}^y_\kappa(A)$ implies $\mathbf{B}^x_\kappa(A)$ and $\mathbf{B}^y_\kappa(A|C)$ implies $\mathbf{B}^x_\kappa(A|C)$.*

5. $\mathbf{B}^\infty_\kappa(A)$ *iff for all n $\mathbf{B}^n_\kappa(A)$; $\mathbf{B}^\infty_\kappa(A|C)$ iff for all n $\mathbf{B}^n_\kappa(A|C)$.*

Proof. Trivial. □

---

[2]Compare Spohn (2012, §5).

[3]κ is a *complete ranking function* over $\wp(W)$ iff $\kappa : \wp(W) \longrightarrow \mathbb{N}_\infty$ is a function, such that (1) $\kappa(W) = 0$, (2) $\kappa(\emptyset) = \infty$ and (3) for all $S \subseteq \wp(W)$, $\kappa(\bigcup S) = \min_{A \in S} \kappa(A)$.

[4]One verifies easily that κ_k is indeed a complete ranking function. Conversely any complete ranking function κ over $\wp(W)$ has a unique ranking mass $k_\kappa(w) = \kappa(\{w\})$.

Ranking Semantics

One may think of a ranking mass as a doxastic ordering source. The zero worlds are the closest worlds, or the best candidates for the actual world. The greater the rank of a world, the less plausible that world is as a candidate for the actual world, the more it is disbelieved or the more the agent has doubts about it. Worlds ranked with $n < \infty$ are within the doxastic modal horizon (or modal base). Worlds with rank ∞ are crazy worlds outside the modal horizon. Although the agent acknowledges their eventual possibility, she disregards them for matters of actual judgements.

The following should then not be astonishing: ranking functions can be used to define $\Box_x$ by x-belief. Similarly, the $>_x$ conditional can be defined by conditional x-belief.

2 Ranking semantics

This section introduces the ranking semantics by turning the preceding remarks into definitions and proves interactions between conditionals and modalities.

We use the *propositional alphabet* $\alpha := \mathcal{V} \cup \{\neg, \wedge\} \cup \{(,)\}$ with *propositional variables* $\mathrm{Var}(\alpha) = \mathcal{V}$. From α and additional modal and conditional operators different languages $\mathcal{L}$ can be generated in the usual way, where generally $\mathrm{Var}(\mathcal{L}) = \mathrm{Var}(\alpha) = \mathcal{V}$.

Definition 1 $\mathfrak{F} = \langle W, (\kappa^w)_{w \in W} \rangle$ *is a **ranked frame** iff (1) $W \neq \emptyset$ is a set of worlds, (2) for each $w \in W$, κ^w is a complete ranking function over $\wp(W)$. $\mathfrak{R} = \langle W, (\kappa^w)_{w \in W}, V \rangle$ is a **ranked model** (based on $\mathfrak{F}$) for $\mathcal{L}$ iff (1), (2) and (3) $V : \mathcal{V} \longrightarrow \wp(W)$ is a valuation.*

A ranked frame (resp. model) is **trivial** if $\kappa^w(A) \in \{0, \infty\}$ for all $A \in \wp(W)$ and all $w \in W$, it is **regular** if all κ^w are regular and it is **global** if there is a ranking function κ s.t. for all w, $\kappa = \kappa^w$.

$\mathcal{L}_{(\Box_x, >_x)_{x \in \mathbb{N}_\infty}}$ is the set of well formed formulas (sentences) obtained by closing $\alpha \cup \{\Box_x : x \in \mathbb{N}_\infty\} \cup \{>_x: x \in \mathbb{N}_\infty\}$ under $\neg, \wedge, \Box_x, >_x$ for $x \in \mathbb{N}_\infty$.[5] We distinguish the *simple modal* and *simple conditional* fragments of type $\mathcal{L}_{\Box_x}$ and $\mathcal{L}_{>_x}$ for fixed $x \in \mathbb{N}_\infty$, the *multi-modal* fragment $\mathcal{L}_{(\Box_n)_{n \in \mathbb{N}}}$, the *multi-conditional* fragment $\mathcal{L}_{(>_n)_{n \in \mathbb{N}}}$ and the *simple interaction* fragment $\mathcal{L}_{\Box_n, >_n}$ (obtained in the obvious way). $\mathcal{L}$ is used to designate any of these languages.

---

[5] As usual, $\vee, \rightarrow$ will be definable and $\Diamond_x \equiv \neg \Box_x \neg$.

225

Definition 2 Let $\mathfrak{R} = \langle W, (\kappa^w)_{w \in W}, V \rangle$ be a ranked model, $w \in W$ and $\varphi \in \mathcal{L}_{(\square_x, >_x)_{x \in \mathbb{N}_\infty}}$. **Truth of φ in $\mathfrak{R}$ in** w, denoted $\langle \mathfrak{R}, w \rangle \vDash \varphi$ (short $w \vDash_\mathfrak{R} \varphi$), is inductively defined as follows, where $\varphi, \psi \in \mathcal{L}_{(\square_x, >_x)_{x \in \mathbb{N}_\infty}}$:

1. $w \vDash_\mathfrak{R} p$ iff $w \in V(p)$, for $p \in \mathcal{V}$,

2. $w \vDash_\mathfrak{R} \neg\varphi$ iff $w \nvDash_\mathfrak{R} \varphi$,[6]

3. $w \vDash_\mathfrak{R} (\varphi \wedge \psi)$ iff $(w \vDash_\mathfrak{R} \varphi$ and $w \vDash_\mathfrak{R} \psi)$,

4. $w \vDash_\mathfrak{R} \square_n \varphi$ iff $\kappa^w([\neg\varphi]) > n$,

5. $w \vDash_\mathfrak{R} \square_\infty \varphi$ iff $\kappa^w([\neg\varphi]) = \infty$,

6. $w \vDash_\mathfrak{R} \varphi >_n \psi$ iff $\kappa^w([\varphi]) = \infty$ or $(\kappa^w)_{[\varphi]}([\neg\psi]) > n$,

7. $w \vDash_\mathfrak{R} \varphi >_\infty \psi$ iff $\kappa^w([\varphi]) = \infty$ or $(\kappa^w)_{[\varphi]}([\neg\psi]) = \infty$.

The sets $[\psi] := [\psi]^\mathfrak{R} = \{w \in W : w \vDash_\mathfrak{R} \psi\}$ are co-inductively defined. I will use the following abbreviations: $\kappa^w(v) = \kappa^w(\{v\})$, $\kappa^w(\varphi) = \kappa^w([\varphi])$, $\kappa^w_\varphi = (\kappa^w)_{[\varphi]}$.

φ is **true in a model** $\mathfrak{R}$, $\mathfrak{R} \vDash \varphi$, iff $w \vDash_\mathfrak{R} \varphi$ in all $w \in W(\mathfrak{R})$. φ is **valid in a class of models** C iff $\mathfrak{R} \vDash \varphi$ for all $\mathfrak{R} \in C$. φ is **valid in a class of frames** F iff for all $\mathfrak{F} \in F$ and all $\mathfrak{R}$ based on $\mathfrak{F}$, $\mathfrak{R} \vDash \varphi$. φ is **valid**, $\vDash \varphi$, iff it is valid in the universal class of all ranked models (equivalently: all ranked frames).

Lemma 2 (interactions)   For all $\varphi, \psi \in \mathcal{L}_{(\square_x, >_x)_{x \in \mathbb{N}_\infty}}$:

1. $w \vDash_\mathfrak{R} \square_x \varphi$ iff $w \vDash_\mathfrak{R} \top >_x \varphi$,

2. $w \vDash_\mathfrak{R} \square_\infty \varphi$ iff $w \vDash_\mathfrak{R} \neg\varphi >_x \bot$,

3. For $y > x$: if $w \vDash_\mathfrak{R} \square_y \varphi$ then $w \vDash_\mathfrak{R} \square_x \varphi$,

4. For $y > x$: if $w \vDash_\mathfrak{R} (\varphi >_y \psi)$ then $w \vDash_\mathfrak{R} (\varphi >_x \psi)$,

5. $w \vDash_\mathfrak{R} \square_\infty \varphi$ iff for all $n \in \mathbb{N}$ $w \vDash_\mathfrak{R} \square_n \varphi$,

6. $w \vDash_\mathfrak{R} \varphi >_\infty \psi$ iff for all $n \in \mathbb{N}$ $w \vDash_\mathfrak{R} \varphi >_n \psi$.

---

[6] $w \nvDash_\mathfrak{R} \varphi$ stands for "not $w \vDash_\mathfrak{R} \varphi$".

Proof. Clear: (1) by Lemma 1.1. (2) since "$\kappa_\varphi^w(\top) > n$ [resp. $\kappa_\varphi^w(\top) = \infty$] or $\kappa^w(\varphi) = \infty$" is equivalent to $\kappa^w(\varphi) = \infty$. (3, 4) by Lemma 1.4 and for (4) the remark that if $\kappa^w([\varphi]) = \infty$, then $w \vDash_\mathfrak{R} \varphi >_z \chi$ anyway. (5,6) by Lemma 1.5 with the same subtlety for (6) as for (4). □

Thus (1) $\square_x$ is definable from $>_x$ and (2) $\square_\infty$ is definable from any $>_x$. Given a conditional $>$ its *inner modality* is $\square \varphi := \top > \varphi$ and its *outer modality* is $\blacksquare \varphi := \neg \varphi > \bot$.[7] Thus $\square_x$ is the inner modality of $>_x$ and $\square_\infty$ its outer modality, i.e., the outer modalities of $>_x$ are all the same and for $x = \infty$, the inner modality is the outer modality. Defining $\Diamond_x = \neg \square_x \neg$, we also obtain:

Corollary 1 (modal hierarchy) *For $m > n$:*

$$\square_\infty \Rightarrow \square_m \Rightarrow \square_n \Rightarrow \square_0 \Rightarrow \Diamond_0 \Rightarrow \Diamond_n \Rightarrow \Diamond_m \Rightarrow \Diamond_\infty.{}^8$$

Proof. By lemma 2.3, duality and the remark that $w \vDash_\mathfrak{R} \neg \square_0 \bot$ and thus $w \vDash_\mathfrak{R} \square_0 \varphi$ implies $w \vDash_\mathfrak{R} \Diamond_0 \varphi$. □

In brief, the $>_x$ are ever stronger conditionals with ever stronger inner modalities $\square_x$, which converge to the limit conditional $>_\infty$ and to the general outer modality $\square_\infty$, respectively. And the inner modalities and their duals generate a hierarchy of modalities which is isomorphic to $\mathbb{Z}$.

Friedman and Halpern (2001) proved completeness of the logic **VD** in $\mathcal{L}_{>0}$ (Lewis' (1991) **V** augmented by the scheme D $\neg(\top > \bot)$). Raidl (ms) showed that the whole Lewisean hierarchy can be reproduced in the general ranking semantics in $\mathcal{L}_{>0}$ (where $\kappa(W) = 0$ is not required), proves completeness results for the impossible antecedent conditionals proposed in Huber (2014, 2015, 2016) and conjectures that the complete logic in $\mathcal{L}_{>n}$ with $n > 0$ is Burgess' system **B** (**V** without CV), augmented by D and the scheme $((\varphi \vee \psi) > \chi) \to ((\varphi > \chi) \vee (\psi > \chi))$. For this reason, I will now focus on the nested modal logic of the ranking semantics.

3 Modal grading, comparatives, multipliers, and reasons

This section states two applications: to the natural language phenomenon of graded modalities, graded comparative modalities and modal multipliers;

---

[7] Another definition is $\blacksquare \varphi := \neg \varphi > \varphi$. Both are equivalent, under some mild assumptions on the conditional logic (ID, CC, RCM in Chellas' notation).

[8] Where M $\Rightarrow$ N for M, N modalities is short for: if $w \vDash_\mathfrak{R} M\varphi$ then $w \vDash_\mathfrak{R} N\varphi$.

and to Spohn's taxonomy of reason relation. This involves some possible extensions of the ranking semantics introduced in §2.

Consider the following example from Kratzer (1991) for graded modality in natural language:

(a) Michl *must* be the murderer.

(b) Michl is *probably* the murderer.

(c) There is a *good possibility* that Michl is the murderer.

(d) Michl *might* be the murderer.

(e) There is a *slight possibility* that Michl is the murderer.

(f) There is a *slight possibility* that Michl is not the murderer.

Kratzer analyses *must* as a very strong necessity and *probably* as a weak necessity, *good possibility* as a strong possibility, *might* as the dual of *must* and thus a very weak possibility and *slight possibility* as a weak possibility. Thus Kratzer's analysis predicts that (a) implies (b) which implies (c) which implies (d), and that (e) also implies (d) and that (f) is incompatible with (a).

Under the following ranking analysis

Operator	Word
$\Box_\infty$	must
$\Box_p$	probably
$\Diamond_g$	good possibility
$\Diamond_s$	slight possibility
$\Diamond_\infty$	might

the above predictions are reproduced, and if $s > g$, then (c) implies (e).[9]

Because of Corollary 1, the ranking semantics has sufficient room to interpret other natural language modalities, such as plausibly, presumably, very good possibility, highly implausible. The ranking semantics can also interpret (graded) modal comparatives, and modal multipliers:

(g) Michl is *more likely* to be the murderer than Jakl.[10]

---

[9] Instead of ∞ we could have chosen any number $M > s, p$.

[10] This stems from Kratzer (1991). For simplicity Kratzer's terminology of 'likely' has been preserved. A better name could be 'plausible' and full adequacy would be obtained if we spoke of 'dubitable'.

Ranking Semantics

(h) Michl is *at least as likely* to be the murderer as Jakl.

(i) Michl and Jakl are *equally likely* to be the murderer.

(j) Michl is *much more likely* to be the murderer than Jakl.

(k) Michl is *way more likely* to be the murderer than Jakl.

(l) It is *twice as likely* that Michl is the murderer than Jakl.

(m) It is *at least twice as likely* that Michl is the murderer than Jakl.

Kratzer (1991) analysis *more likely* as 'being a better possibility' and *at least as likely* as 'being at least as good a possibility' and therefore can analyse (g-i), but has no analysis of (j-m). To obtain an interpretation of (g-m), we enrich Definition 2 by:

8. $w \models_{\Re} \varphi \leq \psi$ iff $\kappa^w(\varphi) \leq \kappa^w(\psi)$,

9. $w \models_{\Re} \varphi < \psi$ iff $\kappa^w(\varphi) < \kappa^w(\psi)$,

10. $w \models_{\Re} \varphi \leq_n \psi$ iff $\kappa^w(\varphi) + n \leq \kappa^w(\psi)$,

11. $w \models_{\Re} \varphi <_n \psi$ iff $\kappa^w(\varphi) + n < \kappa^w(\psi)$,

12. $w \models_{\Re} x\varphi = \psi$ iff $\kappa^w(\varphi)x = \kappa^w(\psi)$,

13. $w \models_{\Re} x\varphi \leq \psi$ iff $\kappa^w(\varphi)x \leq \kappa^w(\psi)$.

Continuing the previous table:

Operator	Word
$\leq$	at least as likely
$<$	more likely
$<_m$	much more likely
$<_w$	way more likely
$x. = .$	x as likely
$x. \leq .$	at least x as likely

The ranking semantics predicts that (g) as (i) imply (h), and that (i) contradicts (g). (l) implies (m). (j) as (k) imply (g), and (m) implies (g) if it is 0-believed that Jakl is not the murderer. And for $w > m$ (k) implies (j). Thus the ranking semantics, in addition to being able to represent graded modalities, can also represent *graded modal comparatives* (g-k), as well as *modal multipliers* (l,m).[11]

We can also extend in another direction and reproduce, in the object language, other ranking-theoretic formalisations of notions from classical epistemology. As an example, let us take Spohn's (2012) analysis of the reason relations. Define $\tau(A) := \kappa(\overline{A}) - \kappa(A)$ and $\tau(A|C) := \kappa(\overline{A}|C) - \kappa(A|C)$. A *is a reason for* C *w.r.t.* κ iff $\tau(C|A) > \tau(C|\overline{A})$. Reasons can be distinguished, here directly in the ranking semantics (cf. Spohn, 2015):

- $w \vDash_{\mathfrak{R}} \varphi >^{\text{sup}} \psi$ iff $\tau(\psi|\varphi) > \tau(\psi|\neg\varphi) > 0$, (supererogatory reason)
- $w \vDash_{\mathfrak{R}} \varphi >^{\text{suf}} \psi$ iff $\tau(\psi|\varphi) > 0 \geq \tau(\psi|\neg\varphi)$, (sufficient reason)
- $w \vDash_{\mathfrak{R}} \varphi >^{\text{nec}} \psi$ iff $\tau(\psi|\varphi) \geq 0 > \tau(\psi|\neg\varphi)$, (necessary reason)
- $w \vDash_{\mathfrak{R}} \varphi >^{\text{ins}} \psi$ iff $0 > \tau(\psi|\varphi) > \tau(\psi|\neg\varphi)$. (insufficient reason)

$>^{\text{sup}}$ and $>^{\text{ins}}$ are *bi-duals* to each other, i.e. $\varphi >^{\text{sup}} \psi$ iff $\neg\varphi >^{\text{ins}} \neg\psi$. The same holds for $>^{\text{suf}}$ and $>^{\text{nec}}$. The necessary and insufficient reason relation can be reproduced in any logic containing a conditional, since in the ranking semantics : $\varphi >^{\text{suf}} \psi$ iff $(\varphi >_0 \psi) \wedge \neg(\neg\varphi >_0 \psi)$. Not so for the other reason relations, since: $\varphi >^{\text{sup}} \psi$ iff $\neg\varphi >_0 \psi$ and if $\neg\varphi >_x \psi$ then $\varphi >_x \psi$ but there is a number $y > 0$ such that $\varphi >_y \psi$ but not $\neg\varphi >_y \psi$. This requires numbers!

In fact, the whole rankyfied epistemology developed by Spohn (2012), such as for example revising or the epistemology of causation, could be reproduced in the object language of ranking semantics. Spohn's theory would then correspond to the non-nested fragment of the logic arising for global ranked models. The additional twist of the ranking semantics is that it has means to analyse the nestings, of belief, conditionals or other operators. Thus auto-epistemic attitudes, such as 'I believe that I believe', are in the object language, and conditions on auto-epistemic attitudes correspond to frame conditions.[12] In other words, the ranking semantics has a neat correspondence theory, which I will now analyse for $\Box_x$.

---

[11] Similar remarks as those made for graded modalities can be made for the conditional, to interpret if-clauses of the form 'if p Modal q'.

[12] The ranking semantics could also be extended to multi-agent frames in order to model the evolution of the common ground, by using revision of ranking functions.

4 Doxastic logic

In this section, I prove completeness for the ranking semantics in the languages $\mathcal{L}_{\Box_x}$ for x fixed, as well as for $\mathcal{L}_{(\Box_n)_{n\in\mathbb{N}}}$.

Given a ranked model $\mathfrak{R}$, consider the following relations:

$$R_n := \{\langle w, v\rangle \in W^2 : \kappa^w(v) \leq n\} \quad , \quad R_\infty := \bigcup_{n\in\mathbb{N}} R_n \qquad (1)$$

For $R \subseteq W^2$ write $R(w) := \{v \in W : wRv\}$. Note that each R_x is a binary serial relation over W and the relations yield nested spheres: $R_n(w) \subseteq R_{n+1}(w) \subseteq R_\infty(w)$. We also have

Lemma 3 $R_n(w) = [\leq n]_{\kappa^w}$. And $R_\infty(w) = [< \infty]_{\kappa^w}$.

Proof. $R_n(w) = \{v \in W : \kappa^w(v) \leq n\} = [\leq n]_{\kappa^w}$. $R_\infty(w) = \{v \in W : \kappa^w(v) < \infty\} = [< \infty]_{\kappa^w}$. □

Given Lemma 1, we can rephrase $\Box_x$ as a Kripke necessity:

Lemma 4 $w \vDash_{\mathfrak{R}} \Box_x \varphi$ iff $R_x(w) \subseteq [\varphi]^{\mathfrak{R}}$.

Proof. For $x = n \in \mathbb{N}$ we have

$$\begin{aligned}
w \vDash_{\mathfrak{R}} \Box_n \varphi \quad &\text{iff} \quad \kappa^w([\neg\varphi]) > n & (\text{Def. } \Box_n) \\
&\text{iff} \quad [\leq n]_{\kappa^w} \subseteq [\varphi]^{\mathfrak{R}} & (\text{Belief Lemma 1.3}) \\
&\text{iff} \quad R_n(w) \subseteq [\varphi]^{\mathfrak{R}} & (\text{Lemma 3})
\end{aligned}$$

The proof is similar for $x = \infty$. □

Let $\mathfrak{M}, \mathfrak{N}$ be two models with same worlds $W(\mathfrak{M}) = W(\mathfrak{N})$ and same valuation $V(\mathfrak{M}) = V(\mathfrak{N})$, but possibly different semantic objects interpreting modalities of some fixed language $\mathcal{L}$. $\mathfrak{M}$ is **faithful to** $\mathfrak{N}$ **in** $\mathcal{L}$ iff for all $\varphi \in \mathcal{L}$ and all $w \in W$, $w, \mathfrak{M} \vDash \varphi$ iff $w, \mathfrak{N} \vDash \varphi$. Given a ranked model $\mathfrak{R} = \langle W, (\kappa^w)_{w\in W}, V\rangle$ restricted to $\mathcal{L}_{\Box_x}$, the x-**model** is $\mathfrak{M}(\mathfrak{R}, x) = \langle W, R_x, V\rangle$. From the previous remarks, it is clear that $\mathfrak{M}(\mathfrak{R}, x)$ is a serial Kripke model, which additionally is faithful to $\mathfrak{R}$ in $\mathcal{L}_{\Box_x}$ (by induction on the complexity of φ, using Lemma 4). Conversely, for a serial Kripke model $\mathfrak{M} = \langle W, R, V\rangle$, its **ranked trivialisation** is $\mathfrak{R}_t(\mathfrak{M}) := \langle W, (\kappa^w)_{w\in W}, V\rangle$, with

$$\kappa^w(v) = \begin{cases} 0 & \text{if } v \in R(w), \\ \infty & \text{else.} \end{cases} \qquad (2)$$

The ranked trivialisation of a serial Kripke model is indeed a trivial ranked model. Thus all necessities are equivalent (since all the R_x are the same) and therefore $\mathfrak{M}(\mathfrak{R}_t(\mathfrak{M}), x) = \mathfrak{M}$. As a consequence: $\mathfrak{R}_t(\mathfrak{M})$ is faithful to $\mathfrak{M}$ in $\mathcal{L}_{\Box_x}$.[13] Soundness and completeness for the ranking semantics in $\mathcal{L}_{\Box_x}$ can now be established by using known results.

φ is a **modal tautology in** $\mathcal{L}$ iff there is a formula $\theta \in \mathcal{L}$ and $p_1, \ldots, p_n \in \mathcal{V}(\mathcal{L})$ and $\psi_1, \ldots, \psi_n \in \mathcal{L}$, such that $\theta = \theta[p_1, \ldots, p_n]$ is a propositional tautology and $\varphi = \theta[\psi_1/p_1, \ldots, \psi_n/p_n]$.

$\Sigma \subseteq \mathcal{L}_\Box$ is a **normal modal logic** iff Σ contains (1) all modal tautologies in $\mathcal{L}_\Box$, (2) all instances of the scheme (K) $\Box(\varphi \to \psi) \to (\Box\varphi \to \Box\psi)$ and Σ is closed under (3) Modus Ponens (if $\varphi, \varphi \to \psi \in \Sigma$, then $\psi \in \Sigma$) and (4) closed under $\Box$-necessitation (if $\varphi \in \Sigma$ then $\Box\varphi \in \Sigma$). **K** is the smallest normal modal logic, **KD** is the smallest normal modal logic containing all instances of the scheme (D) $\neg\Box\bot$ and the normal modal logics **KT**, **K4**, **K5** are defined in the same manner as usually. When we consider $\Box = \Box_x$, we write the schemes S_x and the corresponding normal logics accordingly.

For C a class of models or frames its **theorems** (valid formulas) **in** $\mathcal{L}$ are $\text{Th}_\mathcal{L}(C) := \{\varphi \in \mathcal{L} : C \vDash \varphi\}$. Let $\Sigma \subseteq \mathcal{L}$ and C a class of frames or models. $\Sigma \subseteq \mathcal{L}$ is **sound in** $\mathcal{L}$ for C iff $\Sigma \subseteq \text{Th}_\mathcal{L}(C)$. Σ is **complete in** $\mathcal{L}$ for C iff $\Sigma \supseteq \text{Th}_\mathcal{L}(C)$.

Theorem 1 $\mathbf{K}_x\mathbf{D}_x$ *is sound and complete in* $\mathcal{L}_{\Box_x}$ *for the class of all ranked frames/models.*

Proof. Fix $x \in \mathbb{N}_\infty$ and to avoid switching languages, we locally set $\Box := \Box_x$ and therefore $\mathcal{L}_\Box = \mathcal{L}_{\Box_x}$ as well as $\mathbf{K}_x\mathbf{D}_x = \mathbf{KD}$, etc.
Soundness: Assume $\varphi \in \mathbf{K}_x\mathbf{D}_x$, thus $\varphi \in \mathbf{KD}$. Let $\mathfrak{R}$ be a ranked model in $\mathcal{L}_{\Box_x}$. Then $\mathfrak{M}(\mathfrak{R}, x)$ is a serial Kripke model and $\mathfrak{M}(\mathfrak{R}, x) \vDash \varphi$ (soundness of **KD** for serial Kripke models). Thus $\mathfrak{R} \vDash \varphi$ (faithfulness).
Completeness: Assume $\varphi \notin \mathbf{K}_x\mathbf{D}_x$, thus $\varphi \notin \mathbf{KD}$. Then, for $\Sigma = \mathbf{KD}$, the canonical Kripke model $\mathfrak{M}^\Sigma$ is serial (known fact) and $\mathfrak{M}^\Sigma \nvDash \varphi$ (the canonical model of a normal modal logic determines Σ). Thus $\mathfrak{R}_t(\mathfrak{M}^\Sigma) \nvDash \varphi$ (faithfulness). Hence $\nvDash \varphi$ in the ranking semantics. $\square$

To obtain the weaker logic $\mathbf{K}_x$, we would need to consider **generalised ranking functions** κ, such that $\kappa = \kappa_k$, where k is a **generalised ranking mass** $k : W \longrightarrow \mathbb{N}_\infty$ where it is not required that k has at least one zero.

---

[13] $w \vDash_\mathfrak{M} \varphi$ iff $w \vDash_{\mathfrak{M}(\mathfrak{R}_t(\mathfrak{M}), x)} \varphi$ iff $w \vDash_{\mathfrak{R}_t(\mathfrak{M})} \varphi$ since $\mathfrak{M} = \mathfrak{M}(\mathfrak{R}_t(\mathfrak{M}), x)$ for any x and $\mathfrak{M}(\mathfrak{R}(\mathfrak{M}), x)$ is faithful to $\mathfrak{R}_t(\mathfrak{M})$ by the above remark.

Ranking Semantics

Then $\mathbf{K}_x$ is sound and complete in $\mathcal{L}_{\square_x}$ for the class of generalised ranked frames/models.

Properties of accessibility relations can now be rephrased as properties of ranking functions. We say that the ranked frame property P^x **corresponds to the relation property** P iff : if the ranked frame $\mathfrak{F}$ has P^x then the x-frame has P. As an example:[14]

Definition 3 *Let $\mathfrak{F} = \langle W, (\kappa^w)_{w \in W} \rangle$ be a ranked frame. $\mathfrak{F}$ is*

- *n-reflexive iff for all w, $\kappa^w(w) \leq n$,*

- *n-symmetric iff for all w, v, if $\kappa^w(v) \leq n$ then $\kappa^v(w) \leq n$,*

- *n-transitive iff for all w, v, u $\kappa^w(v) \leq n$ and $\kappa^v(u) \leq n$ imply $\kappa^w(u) \leq n$,*

- *n-euclidean iff for all w, v, u $\kappa^w(v) \leq n$ and $\kappa^w(u) \leq n$ imply $\kappa^v(u) \leq n$,*

- *n-total iff for all w, v $\kappa^w(v) \leq n$,*

- *∞-... iff it is "n-...$[\infty/n]$", the latter being obtained by replacing $\leq n$ by $< \infty$.*

$\mathfrak{R}$ is regular iff it is ∞-total. n-reflexive (or -total) implies $(n+1)$-reflexive (or -total). No such relation exists for the other properties. Note also that: if P^n for all n then P^∞.[15] It is clear that the above properties P^x of ranked frames correspond to the properties P of Kripke frames, in the sense defined above. Conversely, if a serial Kripke frame (or model) has property P then its ranked trivialisation has the property P^x for all $x \in \mathbb{N}_\infty$.

We say that P^x **corresponds** to scheme $(\mathsf{S}^P)_x$ iff the P corresponding to P^x corresponds to axiom scheme S^P in the usual Kripke semantics[16], where S_x is S with all $\square, \Diamond$ being replaced by $\square_x, \Diamond_x$. As an example:

---

[14]Given the definition of "correspondence", the following definition can be proved as a theorem and more generally $hijk$-convergence of a (serial) Kripke relation R can be transformed into a the corresponding x-property of a ranked frame.

[15]This holds for all $hijk$-convergences.

[16]Provided such a correspondence exists, for example for $hijk$-convergences corresponding to Sahlqvist formulas.

Axiom Scheme $(\mathsf{S}^P)_x$	Ranking Property P^x	R_x Property P
(T_x) $\Box_x \varphi \to \varphi$	x-reflexive	reflexive
(B_x) $\varphi \to \Box_x \Diamond_x \varphi$	x-symmetric	symmetric
$(\mathsf{4}_x)$ $\Box_x \varphi \to \Box_x \Box_x \varphi$	x-transitive	transitive
$(\mathsf{5}_x)$ $\Diamond_x \varphi \to \Box_x \Diamond_x \varphi$	x-euclidean	euclidean

Corollary 2 $\mathbf{K}_x \mathbf{D}_x \mathbf{S}_x^{P_1} \ldots \mathbf{S}_x^{P_n}$ *is sound and complete in* $\mathcal{L}_{\Box_x}$ *for the class of ranked frames with properties* $P_1^x \ldots P_n^x$. *And:*

(*) $\mathbf{K}_x \mathbf{T}_x \mathbf{5}_x$ *is sound and complete in* $\mathcal{L}_{\Box_x}$ *for* x*-total ranked frames.*

Proof. Similar proof as above, exploiting (i) the known soundness and completeness results for the corresponding classes of Kripke frames, (ii) the fact that the x-model to $\mathfrak{R}$ is P if $\mathfrak{R}$ is P^x and (iii) the fact that if the canonical model for a normal modal logic with D is P, then its rank-trivialisation is P^∞ and thus P^x for all x. Completeness for (*) is obtained as follows: **KT5** is complete for Kripke frames with R an equivalence relation. Given $\varphi \notin \mathbf{KT5}$, restrict the canonical Kripke model to the equivalence class $[w]$, where $w, \mathfrak{M} \nvDash \varphi$ to obtain the Kripke model $\mathfrak{M}^* := \mathfrak{M}_{[w]}$ which now has a total relation R^* but still has $w, \mathfrak{M}^* \nvDash \varphi$. The ranked-trivialisation of $\mathfrak{M}^*$ is x-total for all x (since $\kappa^w(v) = 0$ for all w and all v) and by faithfulness $w, \mathfrak{R}(\mathfrak{M}^*) \nvDash \varphi$. □

Although $\Box_x$ was constructed as a doxastic necessity, we may remark the following: if we assume ∞-reflexivity, all finite doxastic modalities converge to a knowledge-like (i.e., factive) limit necessity, without being themselves factive. Similarly for regular ranked models, all doxastic modalities converge to a limit knowledge modality (factive, negatively introspectible and positively introspectible). We may also have a doxastic modality $\Box_x$ which already is knowledge like (in some sense) without the previous modalities being knowledge like. Thus different properties of knowledge-likeness can arise at different points in the doxastic hierarchy, some of them will be upper stable, such as factiveness (T), others may be upper- and downwards-unstable, such as introspections (4 or 5).

The correspondence theory for the ranking semantics in $\mathcal{L}_{\Box_x}$ was quite simple. For the multi-modal language $\mathcal{L}_{(\Box_n)_{n \in \mathbb{N}}}$ we will need to define non-trivial ranking analogues of Kripke models.

A **multi-relational Kripke model** is a tuple $\langle W, (R_n)_{n \in \mathbb{N}}, V \rangle$, where each $\langle W, R_n, V \rangle$ is a Kripke model. It is **serial** iff all R_n are serial and

Ranking Semantics

nested iff $R_n(w) \subseteq R_{n+1}(w)$ for all $w \in W$ and all $n \in \mathbb{N}$. A **ranked Kripke model** is a nested and serial multi-relational Kripke model.[17] In a multi-relational Kripke model we define $R_\infty := \bigcup_{n \in \mathbb{N}} R_n$. In a ranked Kripke model, we obtain that R_∞ is serial and $R_n(w) \subseteq R_\infty(w)$. In a multi-relational Kripke model the truh of $\Box_n \varphi$ is defined as usual with respect to the relations R_n of same index.

Definition 4 *Given a ranked model* $\mathfrak{R} = \langle W, (\kappa^w)_{w \in W}, V \rangle$, *its* **kripkefication** *is* $\mathfrak{M}(\mathfrak{R}) := \langle W, (R_n)_{n \in \mathbb{N}}, V \rangle$ *(with R_n the relations in equation 1). Conversely, given a ranked Kripke model* $\mathfrak{M} = \langle W, (R_x)_{x \in \mathbb{N}}, V \rangle$, *its* **rankification** *is* $\mathfrak{R}(\mathfrak{M}) := \langle W, (\kappa^w)_{w \in W}, V \rangle$ *by defining recursively:*

- $[0]_w := R_0(w)$, $[n+1]_w := R_{n+1}(w) \setminus R_n(w)$ *and* $[\infty]_w := W \setminus R_\infty(w)$.

- $\kappa^w = \kappa_{k^w}$ *is induced by setting* $k^w(v) = x$ *iff* $v \in [x]_w$.

One verifies that $\mathfrak{M}(\mathfrak{R})$ is a ranked Kripke model, if $\mathfrak{R}$ is a ranked model (see previous remarks on the R_x) and conversely that $\mathfrak{R}(\mathfrak{M})$ is a ranked model if $\mathfrak{M}$ is a ranked Kripke model (and $\mathfrak{R}(\mathfrak{M})$ is only trivial if $\mathfrak{M}$ is, i.e., $R_\infty = R_0$).[18] Additionally, we have $\mathfrak{M}(\mathfrak{R}(\mathfrak{M})) = \mathfrak{M}$ and this time also $\mathfrak{R}(\mathfrak{M}(\mathfrak{R})) = \mathfrak{R}$, so that the functions $\mathfrak{M}()$ and $\mathfrak{R}()$ are inverses to each other.[19] In the multi-modal language $\mathcal{L}_{(\Box_n)_{n \in \mathbb{N}}}$, we obviously also have that $\mathfrak{M}(\mathfrak{R})$ is faithful to the ranked model $\mathfrak{R}$[20] and $\mathfrak{R}(\mathfrak{M})$ is faithful to the ranked Kripke model $\mathfrak{M}$.[21]

$\Sigma \subseteq \mathcal{L}_{(\Box_n)_{n \in \mathbb{N}}}$ is a **normal multi-modal logic** iff (1) Σ contains all modal tautologies of $\mathcal{L}_{(\Box_n)_{n \in \mathbb{N}}}$ and (2) all instances of $\mathbf{K}_n$ for all $n \in \mathbb{N}$, (3) Σ is closed under Modus Ponens and (4) under $\Box_n$-necessitation for all $n \in \mathbb{N}$. The **ranked (normal multi-modal) logic** $\mathbf{KDN}(\infty)$ is the smallest normal multi-modal logic $\Sigma \subseteq \mathcal{L}_{(\Box_n)_{n \in \mathbb{N}}}$, such that for all $n \in \mathbb{N}$, Σ contains all instances of $\mathbf{D}_n$ and all instances of

- $\Box_{n+1} \varphi \to \Box_n \varphi$ \hfill $(\mathbf{N}_n)$[22]

---

[17] Given nestedness, seriality for R_0 suffices to imply seriality of all R_n.
[18] The $[x]_w$ are well defined (some of them possibly empty), since the R_x are nested. $[0]_w$ is non-empty, since R_0 is serial thus k^w is a ranking mass and κ^w a ranking function.
[19] $v \in R_n(w)$ iff $\kappa^w(v) \leq n$ iff $v \in (R^{\kappa^w})_n(w)$. $\kappa^w(v) \leq n$ iff $v \in R_n(w)$ iff $\kappa_{k^w}(v) \leq x$.
[20] By induction on the complexity of φ, using Lemma 4.
[21] Since $\mathfrak{M}(\mathfrak{R}(\mathfrak{M})) = \mathfrak{M}$ (remark above) and because $\mathfrak{M}(\mathfrak{R}(\mathfrak{M}))$ is faithful to $\mathfrak{R}(\mathfrak{M})$.
[22] N for "nesting". Note that $\mathbf{D}_0$ would have been sufficient.

Eric Raidl

Theorem 2 $\mathbf{KDN}(\infty)$ *is sound and complete in* $\mathcal{L}_{(\Box_n)_{n \in \mathbb{N}}}$ *for ranked models/frames.*

Proof. **Soundness**: Assume $\varphi \in \mathbf{KDN}(\infty)$. Let $\mathfrak{R}$ be an arbitrary ranked model. Then $\mathfrak{M}(\mathfrak{R})$ is a ranked Kripke model (see previous remarks). But then $\mathfrak{M}(\mathfrak{R}) \vDash \varphi$ (the obvious soundness of $\mathbf{KDN}(\infty)$ for ranked Kripke models). Thus $\mathfrak{R} \vDash \varphi$ (faithfullness).

Completeness: Let Σ be a normal multi-modal logic. The canonical model for Σ is $\mathfrak{M}^\Sigma = \langle W^\Sigma, (R_n^\Sigma)_{n \in \mathbb{N}}, V^\Sigma \rangle$, with W^Σ the maximal Σ-consistent theories (which exist by the Lindenbaum Lemma[23]), $v \in R_n^\Sigma(w)$ iff $\{\varphi \in \mathcal{L}_{(\Box_n)_{n \in \mathbb{N}}} : \Box_n \varphi \in w\} \subseteq v$ and $V^\Sigma(p) = \{w \in W^\Sigma : p \in w\}$. The proof then goes along the standard argumentation line. The canonical model of a normal multi-modal logic is a multi-relational Kripke model, i.e., each $\langle W^\Sigma, R_n^\Sigma, V^\Sigma \rangle$ is a Kripke model. The truth lemma holds: for every $\varphi \in \mathcal{L}_{(\Box_n)_{n \in \mathbb{N}}}$, $w, \mathfrak{M}^\Sigma \vDash \varphi$ iff $\varphi \in w$ (by induction). The determination lemma holds (truth lemma and maximal Σ-consistency): if Σ is a normal multi-modal logic, then $\mathfrak{M}^\Sigma \vDash \varphi$ iff $\varphi \in \Sigma$. Additionally:

1. if Σ contains all instances of the schemes $(\mathsf{D}_n)_{n \in \mathbb{N}}$, then each R_n^Σ is serial.

2. if Σ contains all instances of the schemes $(\mathsf{N}_n)_{n \in \mathbb{N}}$, then the R_n^Σ are nested.

(1) is clear. (2) is proven as follows: Suppose $\mathsf{N}_n \subseteq \Sigma$. Let $w \in W^\Sigma$ and suppose $v \in R_n^\Sigma(w)$. Then $\{\varphi : \Box_n \varphi \in w\} \subseteq v$. It now suffices to prove (*): $\{\varphi \in \mathcal{L}_{(\Box_n)_{n \in \mathbb{N}}} : \Box_{n+1} \varphi \in w\} \subseteq \{\varphi \in \mathcal{L}_{(\Box_n)_{n \in \mathbb{N}}} : \Box_n \varphi \in w\}$. Let φ in the first set. Thus $\Box_{n+1} \varphi \in w$. But $\Box_{n+1} \varphi \to \Box_n \varphi \in w$, since w is maximal Σ-consistent and $\mathsf{N}_n \subseteq \Sigma$. Additionally, since Σ is closed under modus ponens, all maximal Σ-consistent theories w are closed under modus ponens. Thus $\Box_n \varphi \in w$ (using the previous). Thus φ is in the second set. By (*) we conclude $\{\varphi : \Box_{n+1} \varphi \in w\} \subseteq v$. Thus $v \in R_{n+1}^\Sigma(w)$. Therefore the canonical model for $\mathbf{KDN}(\infty)$ is a ranked Kripke model.

Assume $\varphi \notin \mathbf{KDN}(\infty)$. Thus $\mathfrak{M}^{\mathbf{KDN}(\infty)} \nvDash \varphi$ (determination). Therefore $\mathfrak{R}(\mathfrak{M}^{\mathbf{KDN}(\infty)}) \nvDash \varphi$ (faithfulness). Hence $\nvDash \varphi$ in the ranking semantics. □

---

[23] As a remark: The axiom of choice is required here only if the set of propositional variables $\mathcal{V} = \mathrm{Var}(\mathcal{L}_{(\Box_n)_{n \in \mathbb{N}}})$ is non-denumerable. Indeed, if $\mathcal{V}$ is denumerable then the alphabet $\beta := \alpha \cup \{\Box_n : n \in \mathbb{N}\}$ is denumerable, as well as its Kleene closure β^* (the set of all finite strings formed from β). But $\mathcal{L}_{(\Box_n)_{n \in \mathbb{N}}} \subseteq \beta^*$. Thus $\mathcal{L}_{(\Box_n)_{n \in \mathbb{N}}}$ is denumerable if β is.

To obtain the system $(\mathbf{K}_x\mathbf{N}_x)_{x\in\mathbb{N}}$, we need to consider generalised ranking functions (cf. the remark made after theorem 2).

Lemma 5 *If a normal multi-modal logic Σ contains $\mathbf{S}_n^P$ then R_n in the canonical model $\mathfrak{M}^\Sigma$ has property P.*

Proof. The proof is the same as for uni-modal Kripke models. □

Corollary 3 $\mathbf{KDN}(\infty)\mathbf{S}_{x_1}^{P_1}\ldots\mathbf{S}_{x_n}^{P_n}$ *is sound and complete in $\mathcal{L}_{(\Box_n)_{n\in\mathbb{N}}}$ for the class of ranked frames having properties $P_1^{x_1},\ldots,P_n^{x_n}$.*

Proof. **Soundness** is clear. **Completeness**: $\mathbf{KDN}(\infty)\mathbf{S}_{x_1}^{P_1}\ldots\mathbf{S}_{x_n}^{P_n}$ is a normal multi-modal logic. Thus its canonical model $\mathfrak{M}$ determines it. Additionally $\mathfrak{M}$ is serial, nested and R_{x_i} has property P_i, for $i=1,\ldots,n$. To $\mathfrak{M}$ there corresponds a faithful canonical ranked model $\mathfrak{R}(\mathfrak{M})$ and the ranking functions have the property $P_i^{x_i}$ for $i=1,\ldots,n$. Assume that $\varphi \notin \mathbf{KDN}(\infty)\mathbf{S}_{x_1}^{P_1}\ldots\mathbf{S}_{x_n}^{P_n}$. Thus $\mathfrak{M}\nvDash\varphi$ (determination). Therefore $\mathfrak{R}(\mathfrak{M})\nvDash\varphi$ (faithfulness) and $\mathfrak{R}(\mathfrak{M})$ has the properties $P_i^{x_i}$ for $i=1,\ldots,n$. Hence $C\nvDash\varphi$ for C the class of ranked models/frames with properties $P_1^{x_1},\ldots,P_n^{x_n}$. □

So much for the doxastic logic of the ranking semantics.

Conclusion

I have extended the ranking semantics introduced in Friedman and Halpern (2001) and discussed some applications to natural language phenomena, such as graded modalities, graded modal comparatives and modal multipliers, as well as to the object language reconstruction of ranking theoretic formalisations (Spohn, 2012) of notions from classical epistemology. Finally, I have proved completeness results for the ranking semantics in the languages $\mathcal{L}_{\Box_x}$ and $\mathcal{L}_{(\Box_n)_{n\in\mathbb{N}}}$. The treated extensions and applications exhaust by no means what is possible in the ranking semantics. But maybe it was enough to underline the beautiful effectiveness of the ranking semantics, where a multitude of modal notions can be treated by a single semantic object which has a well motivated epistemology – the ranking function.

References

Burgess, J. (1981). Quick completeness proofs for some logics of conditionals. *Notre Dame Journal of Formal Logic*, *22*, 76–84.

Friedman, N., & Halpern, J. (2001). Plausibility measures and default reasoning. *Journal of the ACM*, *48*, 648–685.

Halpern, J. Y. (2003). *Reasoning about Uncertainty*. Cambridge: MIT Press.

Huber, F. (2014). New foundations for counterfactuals. *Synthese*, *91*, 2167–2193.

Huber, F. (2015). What should i believe about what would have been the case? *Journal of Philosophical Logic*, *44*, 71–110.

Huber, F. (2017). Why follow the royal rule? *Synthese*, *194*, 1565–1590.

Kratzer, A. (1991). Modality. In A. von Stechow & D. Wunderlich (Eds.), *Semantics: An International Handbook of Contemporary Research* (pp. 639–650). Berlin: de Gruyter.

Lauer, S. (2017). 'I believe' in a ranking-theoretic analysis of 'believe'. In *Proceedings of the Twenty-First Amsterdam Colloquium* (pp. 335–344).

Lewis, D. (1971). Completeness and decidability of three logics of counterfactual conditionals. *Theoria*, *37*, 74–85.

Raidl, E. (ms). Completeness for *counter-doxa* conditionals – using ranking semantics. *Review of Symbolic Logic*.

Spohn, W. (1988). Ordinal conditional functions. A dynamic theory of epistemic states. In W. L. Harper & B. Skyrms (Eds.), *Causation in Decision, Belief Change, and Statistics* (Vol. 2, pp. 105–134). Dordrecht: Kluwer.

Spohn, W. (2012). *The Laws of Belief: Ranking Theory and its Philosophical Applications*. Oxford University Press.

Spohn, W. (2015). Conditionals: A unifying ranking-theoretic perspective. *Philosopher's Imprint*, *15*, 1–30.

Eric Raidl
University of Konstanz
Department of Philosopy
Germany
E-mail: `eric.3.raidl@uni-konstanz.de`

Frege's *Begriffsschrift* and Logicism

JOAN BERTRAN-SAN MILLÁN[1]

Abstract: I put forward a new interpretation of Frege's use of the formal system developed in *Begriffsschrift*, the concept-script. In contrast with the commonly-held view, I argue that this use suggests that he did not articulate a logicist programme in 1879. Two lines of argument support this claim. First, I show that between 1879 and 1882 Frege presented the concept-script of *Begriffsschrift* as a tool for arithmetic, and not as a logical theory from which to deduce arithmetical theorems. Second, I consider Frege's results in *Begriffsschrift* and conclude that they do not imply an endorsement of his later logicist programme.

Keywords: Frege, Logicism, Logistic, *Begriffsschrift*, Concept-script

1 Introduction

It is commonly accepted that Gottlob Frege announced his logicist project in *Begriffsschrift* (1879a). Almost all historical studies agree that there Frege formulated his goal of showing that arithmetic is not an autonomous theory, but is based on logic alone[2]. The formal system developed in *Begriffsschrift*, the concept-script, is thus seen as the first step in the development of Frege's logicist programme.

In this regard, two elements are worth taking into consideration. First of all, without a characterisation of the logicist programme, Frege's endorsement of such a programme in *Begriffsschrift* cannot be adequately addressed. After all, several mathematicians and logicians contemporaneous with Frege agreed in one way or another that arithmetic was reducible to logic; the

---

[1] I am grateful to Calixto Badesa for his careful reading of the paper and helpful suggestions. Thanks to Aldo Filomeno, Juan Luis Gastaldi, Ansten Klev, Ladislav Kvasz and Vera Matarese for comments, and to Michael Pockley for linguistic advice.

The work on this paper was supported by the *Formal Epistemology – the Future Synthesis* grant, in the framework of the *Praemium Academicum* programme of the Czech Academy of Sciences.

[2] See, for instance, (van Heijenoort, 1967, pp. 1–2), (Dummett, 1991, p. 68), (Sluga, 1996, pp. 218–219), (Sullivan, 2004, p. 660) and (Blanchette, 2012, pp. 7–17).

inclusion of Frege in this trend would not mean much. Second, the claim that *Begriffsschrift* is the inaugural step in Frege's logicist project usually involves a noteworthy omission. Right after the publication of *Begriffsschrift*, Frege wrote several papers in which, among other things, he put forward a particular use of the concept-script that was briefly mentioned in *Begriffsschrift*: that this formal system could be applied to scientific disciplines – such as arithmetic or geometry – and improve both their expressive capabilities and their deductive rigour. The 1879–1882 papers were written in the context of a controversy between Frege and Ernst Schröder about their respective formal systems which emerged around the Leibnizian notions of *lingua characterica* and *calculus ratiocinator*[3]. Some historical studies identify Frege's attempt to create a *lingua characterica* with his assumption of the logicist thesis[4].

This paper is in two parts. In the first I shall characterise such an application of the concept-script: I shall consider the particularities of its language, the changes in interpretation of its symbols, and its deductions. In the second part, after a presentation of the basic elements of an articulated logicist programme, I shall claim, on the one hand, that the application of the concept-script is not compatible with such a programme and, on the other, that the results of *Begriffsschrift* do not show any endorsement of the logicist thesis.

2 Logistic: instrumental use of the concept-script

In the Preface to *Begriffsschrift* Frege associated the construction of the concept-script with a twofold goal. On the one hand, this formal system is a means to establish rigorous foundations for some propositions which are relevant in arithmetic and demonstrate that their proofs do not have to appeal to intuition (1879a, p. 104). On the other hand, Frege aimed at the construction of a formal structure fit to complement scientific languages, one that is capable of being used as an aid for the rigourisation of scientific proofs and the processes of concept formation (1879a, p. 106).

The result of the application of the concept-script to a given scientific discipline is a hybrid system which may be called *logistic*[5]. The use of the

---

[3] According to Frege, the fact that the application of the concept-script to a scientific discipline is capable of expressing content in a rigorous and unambiguous way is essential for considering the concept-script to be the basis of a realisation of a *lingua characterica*.

[4] See (Sluga, 1987, pp. 90–92), (Peckhaus, 2004, pp. 9–10) and (Korte, 2010, pp. 291–292).

[5] My use of the term 'logistic' is non-standard; it is introduced as a means to refer to Frege's instrumental use of logic (cfr. Church 1956, pp. 47–58). The notion of logistic has been traditionally opposed to the abstract use of logic (i.e., to symbolic logic). In his monograph *A*

Frege's *Begriffsschrift* and Logicism

concept-script as the basis of a logistic system by Frege is a central element in 'Anwendungen der Begriffsschrift' (1879b), 'Booles rechnende Logik und die Begriffsschrift' (1880) and 'Über den Zweck der Begriffsschrift' (1882).

3 Logistic: language

The language of the concept-script of *Begriffsschrift* is – to the contemporary eye – peculiar. First, in this work, Frege did not provide a definition of the notion of atomic formula. The most basic expression of the concept-script, '$f(a)$', can be interpreted in different ways. This is due to the generality expressed by the letters occurring in '$f(a)$'. In fact, the letter 'a' can be interpreted, depending on the context, as a sentential variable, as an individual variable or as a predicate variable[6]; the letter 'f' can also be subject to multiple readings that fit with those of 'a'. Therefore, it cannot be determined beforehand whether '$f(a)$' is an atomic formula or not.

Second, besides the letters, the language of the concept-script lacks non-logical symbols. In particular, there are no individual constants or predicate symbols in the language of the concept-script. Only the logical symbols – judgement and content strokes, connectives and the generality symbol – have a unique possible reading. Consequently, disregarding the propositional fragment of the concept-script, it is not possible to express any definite meaning by means of a concept-script formula; for instance, it is not possible to univocally express in a single formula of the concept-script that an object has some property[7].

In contrast, scientific discourses have non-logical symbols that refer to specific objects and relations. By means of them, it is possible to build atomic formulas. However, these discourses typically lack the formal resources

---

Survey on Symbolic Logic (1918) Lewis presented a characterisation of logistic that fits with Frege's instrumental use of the concept-script:

> "[L]ogistic" is commonly used to denote symbolic logic together with the application of its methods to other symbolic procedures. Logistic may be defined as *the science which deals with types of order as such.* (...) Its subject matter is not confined to logic. (Lewis, 1918, p. 3)

[6] As illustrations of the possible interpretations of the letters 'a' and 'c', see the derivations of propositions (89), (92) and (77) in *Begriffsschrift*.

[7] Throughout Chapter III of *Begriffsschrift*, Frege employs the letter 'f' as a parameter for procedures. In this context, an expression such as '$f(x, y)$' would only be interpreted as an atomic formula: y is a result of the application of the procedure f to x. I shall consider this letter in Section 6.

needed to express the logical relations that bind the atomic formulas together. They thus rely on natural language as a means to refer to complex notions or to define new concepts. Moreover, scientific discourses need the assistance of natural language in the construction of proofs (Frege, 1880, p. 13). In order to avoid the ambiguities and inaccuracies produced by the use of natural language, Frege proposed using the concept-script as the basis of what I term a system of logistic:

> What we have to do now, in order to produce a more adequate solution [than Boolean logic], is to supplement the signs of mathematics with a formal element, since it would be inappropriate to leave the signs we already have unused (...). Thus, the problem arises of devising signs for logical relations that are suitable for incorporation into the formula-language of mathematics, and in this way forming – at least for a certain domain – a complete concept-script. This is where my booklet [*Begriffsschrift*] comes in. (Frege, 1880, pp. 13–14)

Essentially, a system of logistic is the result of adding to the language of the concept-script the non-logical symbols of a scientific discipline (and thus, a means for building atomic formulas) and the construction of a calculus based upon the axiomatic system of the concept-script with the addition of some truths pertaining to the discipline in question. It is thus possible to understand why, in the exposition of the language of the concept-script of *Begriffsschrift*, Frege introduced neither a single non-logical symbol (i.e., individual constant, predicate or relation symbol) nor defined the notion of an atomic formula. This omission should not be seen as an epochal slip, since it is perfectly coherent if Frege's aim is observed. Departing from the atomic formulas of a given discipline, i.e., from those statements that contain no connectives or quantification, an expression in a system of logistic is built by using the logical symbols of the concept-script.

On several occasions Frege exemplified the application of the language of the concept-script to a particular discipline. In 'Booles rechnende Logik und die Begriffsschrift' (1880), he discussed a translation into logistic of an informal arithmetical statement:

> If every square root of 4 is a 4th root of m, then m must be 16.
> The expression

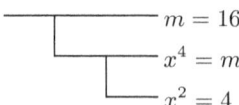

does not correspond to the sentence, and is even false (...); for we may substitute numbers for x and m which falsify this content. (Frege, 1880, p. 18)

Frege wanted to highlight that an adequate logical analysis is an essential step when statements are to be formally expressed. The two different ways in which the concept-script can express generality render possible the appropriate symbolisation of the statement, "If every square root of 4 is a 4th root of m, then m must be 16". The resulting logistic expression is the following:

$$\vdash \begin{array}{l} m = 16 \\ \mathfrak{a}^4 = m \\ \mathfrak{a}^2 = 4. \end{array}$$

We can see in this example that Frege did not aim at symbolisation as it could be understood nowadays. In particular, he did not replace the non-logical symbols of the language of a given discipline with non-logical constants. In this sense, a system of logistic is not properly a formalised theory – such as Peano arithmetic. On the contrary, Frege just wanted to adequately render the formal complexity of the statements of a discipline by means of logical symbols while, at the same time, retaining the specific meaning these statements express. All atomic expressions, and the non-logical symbols occurring in them – symbols that denote, for instance, numbers and numerical operations – remain intact. This means that these non-logical symbols are not reinterpretable; since they do not acquire different meanings as do the individual constants and predicate symbols of first-order languages, the non-logical symbols in logistic should be considered canonical names.

4 Logistic: semantics

Scientific disciplines that are part of a system of logistic have a specific field of application which determines a domain of entities. The non-logical symbols of these disciplines denote objects in this domain, or relations and properties over objects in the domain. By means of these symbols and the presence of such a domain, the expressions of a system of logistic – unlike those of the isolated concept-script – acquire specific meanings and, in this sense, can be considered to be expressions of, as Frege put it, a *complete* concept-script.

As stated above, a concept-script letter can be interpreted in different ways. Only when the formula in which a letter occurs is used in a specific context – paradigmatically in a derivation – can the relevant interpretation of the letter be determined. When concept-script letters occur in a logistic expression, their generality is restricted in two different ways. As an illustration, consider Proposition (52) of *Begriffsschrift*:

$$\vdash \begin{array}{l} f(d) \\ f(c) \\ (c \equiv d) \end{array} \qquad (52)$$

This proposition can be interpreted in at least two different ways: propositionally, or as a first-order logic formula.

Consider now the following instances of Proposition (52):

$$\vdash \begin{array}{l} 2+1 > 0 \\ c > 0 \\ (c \equiv 2+1) \end{array} \quad \text{(A)} \qquad \vdash \begin{array}{l} (3^2 > 1 \equiv n > 1) \\ (c \equiv n > 1) \\ (c \equiv 3^2 > 1) \end{array} \quad \text{(B)}$$

Both formulas could belong to the same system of logistic (in this case, the application of the concept-script to arithmetic). Note that, on the one hand, both (A) and (B) can only be interpreted in one way: the letter 'c' thus loses the multiple readings it can have in (52). In (A) 'c' is interpreted as an individual variable and, in contrast, it can only be propositionally interpreted in (B). On the other hand, the two incompatible interpretations of 'c' in (A) and (B) are also limited by the arithmetical language and the interpretation of arithmetical symbols. In particular, only numerical terms can take the place of the occurrences of 'c' in (A), whilst the appropriate instances of 'c' in (B) are arithmetical formulas.

Quantifiers in the applied concept-script are also adapted to the presence of a specific domain of entities and thus their meaning is restricted. All quantified letters still express generality, yet this generality is limited not only by the syntactic conditions imposed by the concept-script but also by the semantic restrictions laid down by the discipline.

This circumstance is apparent in 'Booles rechnende Logik und die Begriffsschrift' (1880). In this paper from Frege's *Nachlaß* the following example of definition, rendered as a logistic expression, can be found:

Frege's *Begriffsschrift* and Logicism

The real function $\Phi(x)$ is continuous at $x = A$; that is, given any positive non-zero number $\mathfrak{n}$, *there is* a positive non-zero $\mathfrak{g}$ such that any number $\mathfrak{d}$ lying between $+\mathfrak{g}$ and $-\mathfrak{g}$ satisfies the inequality $-\mathfrak{n} \leqq \Phi(A + \mathfrak{d}) - \Phi(A) \leqq \mathfrak{n}$

$$\vdash\!\!\!-\overset{\mathfrak{n}}{\smile}\!\!\!\top\!\!\overset{\mathfrak{g}}{\smile}\!\!\!\top\!\!\overset{\mathfrak{d}}{\smile}\!\!\!\begin{array}{l} -\mathfrak{n} \leqq \Phi(A+\mathfrak{d}) - \Phi(A) \leqq \mathfrak{n} \\ -\mathfrak{g} \leqq \mathfrak{d} \leqq \mathfrak{g} \\ \mathfrak{g} > 0 \\ \mathfrak{n} > 0 \end{array}$$

I have assumed here that the signs $<, >, \leqq$ mark the expressions they stand between as real numbers. (Frege, 1880, p. 24)

This is an example of the application of the language of the concept-script to mathematical analysis. In this context the quantified letters '$\mathfrak{n}$', '$\mathfrak{g}$' and '$\mathfrak{d}$' are not individual variables that take values over an unrestricted domain, but letters that express generality exclusively over real numbers.

As a consequence, a statement in logistic – for instance, one which can be obtained from the previous example – is not logical, i.e., it is not a logical truth whose validity does not depend on a specific interpretation. A statement in logistic is intended to be true *only* in the discipline to which the concept-script is applied. In his examples, Frege did not only restrict the interpretation of the quantifiers but also avoided any reinterpretation of the non-logical symbols, which would be the usual practice nowadays. For instance, in the last example '>' is not taken to be a logical relation – applicable to any pair of objects – but a specific relation in mathematical analysis that is applicable only to real numbers.

5 Logistic: derivations

A system of logistic is not only the result of putting together the basic laws of the concept-script and a set of formulas that express facts about a discipline. Such a system would not be usable in derivations, since the basic laws of the concept-script do not contain any non-logical constant and thus their content is not connected with the domain of the discipline.

Actually, no proposition of the concept-script can directly participate in a derivation of a system of logistic. Only applications of the logical laws of the concept-script can be fruitfully used in a derivation that also involves formulas with a specific and unique reading. In order to render compatible the logical laws of the concept-script and the formulas of a system of logistic, some or all of the letters of those logical laws have to be replaced with the appropriate expressions containing those symbols found in the discipline. In other words, by means of substitutions, an expression of logistic is obtained as an application of a logical law of the concept-script. Through this process, a single interpretation of a logical law is fixed. This is why substitutions are essential in derivations in a system of logistic.

Therefore, only an application of the formulas of the concept-script is involved in derivations in logistic. This was shown to be the case by Frege in an example of a derivation he provided in 'Booles rechnende Logik und die Begriffsschrift' (1880, pp. 27–32); he first introduced the propositions of the concept-script that needed to be incorporated in the proof and he then indicated the substitutions that allowed the use of the appropriate instances of the logical laws in the derivation. In particular:

> In addition we need the formula (4) which is introduced as (96) on p. 71 of the *Begriffsschrift*. It means: if y follows x in the f-series, then every result of applying the procedure f to y follows x in the f-series[8]:
>
> $$\vdash \begin{array}{l} \overset{\gamma}{\underset{\beta}{}} f(x_\gamma, z_\beta) \\ f(y, z) \\ \overset{\gamma}{\underset{\beta}{}} f(x_\gamma, y_\beta) \end{array} \qquad (4)$$
>
> (....) [W]e substitute $x + a = y$ for $f(x, y)$, 0 for x, $(n + b)$ for y and $(n + m)$ for z in (4), giving us (6)[9]:
>
> $$\vdash \begin{array}{l} \overset{\gamma}{\underset{\beta}{}} (0_\gamma + a = (n+m)_\beta) \\ (n + b) + a = n + m \\ \overset{\gamma}{\underset{\beta}{}} (0_\gamma + a = (n+b)_\beta) \end{array} \qquad (6)$$

(Frege, 1880, pp. 28–29)

Taken in isolation, there are no deductions with premises in the calculus of the concept-script: logical laws are obtained exclusively from basic laws and

---

[8] In this context, it can be of benefit to consider the successor function as an example of a procedure and the numerical order $<$ as an example of the series resulting from the application of the successor; accordingly, '$\overset{\gamma}{\underset{\beta}{}}(n_\gamma + 1 = m_\beta)$' means that '$n < m$'.

[9] In this second formula, the procedure $x + a = y$ is defined by the operation $+a$: it relates a number with the result of adding a to it. Then the series associated to this procedure and starting with 0 corresponds to an ordering of the multiples of a.

other logical laws. However, a system of logistic does have a discipline's set of formulas which can be used as premises. In the above derivation in logistic, Frege set the goal of proving the theorem that "the sum of two multiples of a number is in its turn a multiple of that number" (1880, pp. 27–32). As a means to attaining this goal, he used two arithmetical laws as premises. Frege even distinguished these arithmetical laws from the "theorems of pure thought" he needed as logical laws in the proof – one of them, the formula (4) I have just considered. Accordingly, when a set of formulas of a discipline is available, then a deduction with premises can be considered in the calculus of the concept-script. As Frege clearly showed in this example, the premises used in a proof of logistic do not need to be axioms of the discipline. It is enough to isolate a set of formulas that are relevant in a given context, just as Frege did.

Once the syntactic structure of all formulas that take part in a derivation has been rendered uniform by means of substitutions, the inference rules of the concept-script can be used normally. After all, the distinction between function and argument, on which the inference rules of the concept-script of *Begriffsschrift* are based, is flexible enough to be applied to the expressions of any regimented language. Since the expressions of a system of logistic are constructed according to the syntactic rules of the concept-script – and, in particular, using its logical symbols – the application of inference rules is straightforward.

6 Concept-script and logicism

In this last section, after the consideration of how Frege intended to use the concept-script, I address the claim that he started his logicist programme in *Begriffsschrift*. The following passage of *Begriffsschrift*'s Preface is often cited as evidence for this claim:

> (...) I had first to test how far one could get in arithmetic by means of logical deductions alone, supported only by the laws of thought, which transcend all particulars. The procedure in this effort was this: I sought first to reduce the concept of ordering-in-a-sequence to the notion of *logical* ordering, in order to advance from here to the concept of number. (Frege, 1879a, p. 104)

The main goal in the creation of the concept-script in *Begriffsschrift* is usually considered to be the answer to the question of how far one could get in arithmetic by means of logical deductions alone. Therefore, from this perspective, the concept-script was created by Frege with the logicist programme in mind.

In light of Frege's exposition in 1879–1882 concerning the use of the concept-script, a reading of the last paragraph of the Preface to *Begriffsschrift* suggests a different diagnosis:

> Arithmetic, as I said at the beginning, was the starting point of the train of thought which led me to my "concept-script". I intend, therefore, to apply it to this science first, trying to analyse its concepts further and provide a deeper foundation for its theorems. For the present, I have presented in the third chapter some things which move in that direction. Further pursuit of the suggested course – the elucidation of the concepts of number, magnitude, and so forth – is to be the subject of further investigations which I shall produce immediately after this book. (Frege, 1879a, p. 107)

Frege's account fits with the explained instrumental use of the concept-script. In fact, Frege hinted in this passage at the two main elements that have been explained: on the one hand, the combination of the logical symbols and letters of the concept-script with the atomic statements of arithmetic to amend the shortcomings of the latter with regard to the process of concept formation; and, on the other hand, the reconstruction of arithmetical proofs by supplementing them with the formal resources of the concept-script calculus.

In order to assess whether this instrumental use of the concept-script is consistent with an endorsement of logicism, the basic elements of Frege's project of the reduction of arithmetic into logic should be expounded. A successful articulation of the logicist project demands the following[10]:

1. A specification of what is understood by logic, which includes a clarification of what constitutes a logical notion and a logical law;

---

[10]Frege never used the term 'logicism' (i.e., the German '*Logizismus*'); logicism was first attributed to Frege as a foundational thesis of the reduction of mathematics into logic by Carnap (1931, p. 91). Interestingly, Carnap mentioned *Grundlagen der Arithmetik* (Frege, 1884) as the first work to advocate logicism. I am indebted to Marco Panza for this historical remark.

My account of Frege's logicist project does not mean to be exhaustive or definitive. The only goal of such an account is to serve as a minimal and clear model that enables the evaluation of Frege's position between 1879 and 1882. It is based on (Bays, 2000, pp. 415–416) and Badesa's unpublished material. Alternative approaches to the nature of Frege's logicism can be found in (Parsons, 1965), (Benacerraf, 1981), (Demopoulos & Clark, 2005), (Rayo, 2005) and (Kremer, 2006), among others. Concerning a historical reconstruction of the sources of Frege's logicism, see (Reck, 2013).

2. A logical calculus, composed of a determined set of basic laws and a limited and well-specified set of inference rules;

3. A justification that all basic concepts of arithmetic are logical notions, i.e., that they can be defined explicitly by means of logical notions[11];

4. A proof of all arithmetical laws in terms of the definitions obtained in (3) and the elements of the calculus specified in (2).

There is no trace of such an articulation of the logicist thesis in the papers written right after the publication of *Begriffsschrift* in 1879. In these papers Frege did not provide a single explicit definition of an arithmetical notion by means of logical notions; on the contrary, arithmetical concepts were defined using simpler arithmetical concepts. Therefore, arithmetic retains its basic notions and consequently its domain of specific objects, relations and operations. According to this, Frege maintained both the non-logical symbols of arithmetic and the restrictive interpretation of letters and quantifiers in complementing arithmetic with the concept-script, precisely with the aim of producing statements that refer to facts about numbers and the operations between them. Moreover, in the sole derivation of an arithmetical law that can be found in the papers written between 1879–1882 (the previously mentioned proof of the theorem that "the sum of two multiples of a number is in its turn a multiple of that number" (1880, pp. 27–32)) Frege used two arithmetical laws as premises. This is an implicit acknowledgement that the truth of the theorem to be proven is not founded on logic alone.

The results of Chapter III of *Begriffsschrift* raise another question regarding Frege's commitment to logicism in 1879. In this chapter he presented a peculiar example of the instrumental use of the concept-script: this formal system is used to obtain what Frege called "some propositions about sequences" (1879a, §23, p. 167). This is possible by providing symbols with a fixed, albeit abstract interpretation, such as, for instance, 'f' and 'F', which express generality over procedures and properties, respectively. Frege even introduced new letters, such as 'x', 'y', 'z' or 'm', which – unlike the regular letters of the concept-script taken in isolation – have a stable domain of interpretation: they express generality over objects. On this basis, Frege could define the notions of hereditary property, weak and strong ancestral, and single-valued (*eindeutig*) procedure, and derive some theorems that state basic properties of these notions, in particular, the principle of mathematical induction, i.e., Proposition (81) (1879a, §27, pp. 176–177).

---

[11] On the role played by explicit definitions in (3), see (Klev, 2017, pp. 342–344).

Joan Bertran-San Millán

Frege had a calculus in *Begriffsschrift*, which allowed him to clearly specify what he understood by a logical method of proof and thus fulfil demand (2). However, the results of *Begriffsschrift* do not even show a partial commitment to logicism. Firstly, no justification that the basic notions employed in Chapter III are logical can be found in *Begriffsschrift*. Specifically, the notions of hereditary property or strong ancestral rely on the notion of procedure, which is introduced without any clarification: Frege merely translated into natural language a formula in which a binary function letter 'f' occurs[12]. Secondly, since Frege did not clarify what he understood by logic, he failed to justify the claim that the resulting theorems in Chapter III are logical laws in the sense specified in demand (4)[13].

All in all, the concept-script could eventually be tied to the development of the logicist thesis, but to serve as its vehicle was not its sole function. Not only did Frege explicitly intend to use this formal system in ways which are incompatible with a full assumption of logicism, but also only a few of the elements needed to fully articulate this thesis were present in *Begriffsschrift*, in which the concept-script was first introduced.

Pertaining to this, it is noteworthy that after the failure of Frege's logicist project, he kept using the concept-script, albeit in the more elaborate *Grundgesetze* version. According to Carnap's student notes (Reck & Awodey, 2004), Frege maintained the basic components of the concept-script – excluding the notion of value range and Basic Law (V) – and used this formal system instrumentally, as a tool for the rigourisation of arithmetic.

Frege might have had intuitions concerning the logical nature of arithmetical truths in 1879–1882, and even before this period[14]. Given this, his

---

[12]The fact that Frege considered these notions to be within the realm of pure thought does not entail that he substantiated his position. In fact, I do not want to claim here that Frege ever successfully characterised the logical nature of those notions he took as basic. However, he explicitly addressed this matter from *Grundlagen* onwards – where the notions of object and concept are essential – and restricted their attributes to the ones he considered to be undeniably logical (see especially Frege, 1884, §27, p. 37, fn and Frege, 1884, §74, p. 87). See also the table of contents of the unfinished 'Logik' (1882–1891, p. 1).

[13]In *Begriffsschrift* Frege hinted at a basic idea of what he understood by logical laws, the "laws upon which all knowledge rests" and "transcend all particulars" (1879a, pp. 103–104); he pointed to their maximum generality (see also Frege, 1879a, p. 167). Frege first characterised in some detail the notion of logical law in 'Logik' (1882–1891, pp. 3–7) and further developed his position in other works and considered, at length, the justification of the laws of logic in *Grundgesetze der Arithmetik* (1893, pp. xiv–xix).

[14]In the list of theses Frege provided when he defended his dissertation "Über eine geometrische Darstellung der imaginären Gebilde in der Ebene" (1873), he included as the third thesis, "Number is not something originally given [*ursprünglich Gegebenes*], but can be defined".

reference to the reduction of the concept of number in the Preface to *Begriffsschrift* can be understood. However, Frege did not articulate these alleged intuitions either philosophically or formally. Besides, as a result of my analysis, I conclude that he could not have defended them as a programmatic goal without contradicting key features of his factual use of the concept-script.

References

Bays, T. (2000). The fruits of logicism. *Notre Dame Journal of Formal Logic*, *41*, 415–421.

Benacerraf, P. (1981). Frege: The last logicist. *Midwest Studies in Philosophy*, *6*, 17–36.

Blanchette, P. (2012). *Frege's Conception of Logic*. Oxford: Oxford University Press.

Carnap, R. (1931). Die logizistische Grundlegung der Mathematik. *Erkenntnis*, *2*, 91–105.

Church, A. (1956). *Introduction to Mathematical Logic* (Vol. I). Princeton: Princeton University Press.

Demopoulos, W., & Clark, P. (2005). The logicism of Frege, Dedekind and Russell. In S. Shapiro (Ed.), *The Oxford Handbook of Philosophy of Mathematics and Logic* (pp. 129–165). Oxford: Oxford University Press.

The complete list of theses can be found in (Kreiser, 2001, p. 123). I am indebted to Ansten Klev for this quotation.

In 'Booles rechnende Logik und die Begriffsschrift' (1880), Frege considered the possibility of using – as algebraic logicians did – the symbols of arithmetical operations to render logical relations. While examining this logical use of arithmetical symbols, Frege hinted at what could be seen as a logicist position:

> Anyone demanding the closest possible agreement between the relations of the signs and the relations of the things themselves will always feel it to be back to front when logic, whose concern is correct thinking and *which is also the foundation of arithmetic*, borrows its signs from arithmetic. To such a person it will seem more appropriate to develop for logic its own signs, derived from the nature of logic itself; we can then go on to use them throughout the other sciences wherever it is a question of preserving the formal validity of a chain of inference. (Frege, 1880, p. 12, author's emphasis)

Note that, besides this remark, Frege did not further develop his intuition regarding the nature of logic in 'Booles rechnende Logik und die Begriffsschrift' (1880). Actually, as I have showed, a significative part of this paper consists in a defense of the possibility of using the concept-script instrumentally.

Dummett, M. (1991). *Frege: Philosophy of Mathematics.* Harvard: Harvard University Press.

Frege, G. (1873). *Über eine geometrische Darstellung der imaginären Gebilde in der Ebene* (Unpublished doctoral dissertation). Philosophischen Fakultät zu Göttingen, Göttingen. (Published in Jena: A. Neuenhahn)

Frege, G. (1879a). *Begriffsschrift, eine der arithmetischen nachgebildete Formelsprache des reinen Denkens.* Halle: Louis Nebert. (Reedition in (Frege, 1964, pp. 1–88). English translation in (Frege, 1972, pp. 101–203))

Frege, G. (1879b). *Anwendungen der Begriffsschrift.* (Lecture at the January 24, 1879 meeting of *Jenaischen Gesellschaft für Medizin und Naturwissenschaft.* Published in 1879 in *Jenaische Zeitschrift für Naturwissenschaft, 13,* pp. 29–33)

Frege, G. (1880, –1881). *Booles rechnende Logik und die Begriffsschrift.* (Originally unpublished. Edition in (Frege, 1969, pp. 9–52). English translation in (Frege, 1979, pp. 9–46))

Frege, G. (1882). *Über den Zweck der Begriffsschrift.* (Lecture at the January 27, 1882 meeting of *Jenaischen Gesellschaft für Medizin und Naturwissenschaft.* Published in 1882 in *Jenaische Zeitschrift für Naturwissenschaft, 16*, pp. 1–10. English translation in (Frege, 1972, pp. 90–100))

Frege, G. (1882–1891). *Logik.* (Originally unpublished. Edition in (Frege, 1969, pp. 1–8). English translation in (Frege, 1979, pp. 1–8))

Frege, G. (1884). *Die Grundlagen der Arithmetik: eine logisch-matematische Untersuchung über den Begriff der Zahl.* Breslau: Wilhelm Koebner.

Frege, G. (1893). *Grundgesetze der Arithmetik. Begriffsschriftlich abgeleitet* (Vol. I). Jena: Hermann Pohle.

Frege, G. (1964). *Begriffsschrift und andere Aufsätze* (I. Angelelli, Ed.). Hildesheim: Georg Olms. (English translation in (Frege, 1972))

Frege, G. (1969). *Nachgelassene Schriften* (H. Hermes, F. Kambartel, & F. Kaulbach, Eds.). Hamburg: Felix Meiner. (English translation in (Frege, 1979))

Frege, G. (1972). *Conceptual Notation and Related Articles* (T. W. Bynum, Ed.). Oxford: Clarendon Press.

Frege, G. (1979). *Posthumous Writings* (H. Hermes, F. Kambartel, & F. Kaulbach, Eds.). Chicago: University of Chicago Press.

Klev, A. (2017). Dedekind's logicism. *Philosophia Mathematica, 25,* 341–368.

Korte, T. (2010). Frege's *Begriffsschrift* as *lingua characterica*. *Synthese*, *174*, 283–294.
Kreiser, L. (2001). *Gottlob Frege. Leben - Werk - Zeit*. Hamburg: Felix Meiner.
Kremer, M. (2006). Logicist responses to Kant: (early) Frege and (early) Russell. *Philosophical Topics*, *34*, 163–188.
Lewis, C. I. (1918). *A Survey of Symbolic Logic*. Berkeley: University of California Press.
Parsons, C. (1965). Frege's theory of number. In M. Black (Ed.), *Philosophy in America* (pp. 180–203). Ithaca: Cornell University Press.
Peckhaus, V. (2004). Calculus ratiocinator versus characteristica universalis? The two traditions in logic, revisited. *History and Philosophy of Logic*, *25*, 3–14.
Rayo, A. (2005). Logicism reconsidered. In S. Shapiro (Ed.), *The Oxford Handbook of Philosophy of Mathematics and Logic* (pp. 203–235). Oxford: Oxford University Press.
Reck, E. H. (2013). Frege, Dedekind, and the origins of logicism. *History and Philosophy of Logic*, *34*, 242–265.
Reck, E. H., & Awodey, S. (Eds.). (2004). *Frege's Lectures on Logic: Carnap's Student Notes, 1910–1914*. Chicago: Open Court.
Shapiro, S. (Ed.). (2005). *The Oxford Handbook of Philosophy of Mathematics and Logic*. Oxford: Oxford University Press.
Sluga, H. (1987). Frege against the booleans. *Notre Dame Journal of Formal Logic*, *28*, 80–98.
Sluga, H. (1996). Frege on meaning. *Ratio*, *9*, 209–226.
Sullivan, P. (2004). Frege's logic. In D. M. Gabbay & J. Woods (Eds.), *Handbook of the History of Logic* (Vol. 3: The Rise of Modern Logic: From Leibniz to Frege, pp. 659–750). Amsterdam: Elsevier North Holland.
van Heijenoort, J. (Ed.). (1967). *From Frege to Gödel, a Source Book in Mathematical Thought*. Cambridge: Harvard University Press.

Joan Bertran-San Millán
The Czech Academy of Sciences, Institute of Philosophy
The Czech Republic
E-mail: sanmillan@flu.cas.cz

Reasoning About Fiction

TOM SCHOONEN AND FRANCESCO BERTO[1]

Abstract: In this paper we provide a suggestion on how to model reasoning about fiction. We will argue that three things happen: (1) one (sequentially) takes on board the explicit content of the fiction; (2) one imports background beliefs; this is represented by the most plausible worlds after an update triggered by the explicit content; and (3) from these most plausible worlds, an agent reasons on alternative courses of action within the fiction. We suggest to model (1) and (2) by means of a *plausibility* ordering on worlds after having updated with the explicit fictive content. Finally, we use an *inferential* ordering on worlds to model (3).

Keywords: Truth in Fiction, Impossible Worlds, Counterfactual Reasoning, Dynamic Epistemic Logic, Plausibility Ordering

Introduction

People often reason about fiction. For example, the truth of this sentence is hotly debated amongst *Lord of the Rings* fans:

(1) If Frodo had *flown* to Mount Doom on an eagle, rather than *walking* there in order to destroy the ring, there would have been a quick and easy victory in the war with Sauron.

On the one hand people argue that eagles are shown to be strong enough to carry men in the stories of Tolkien, and capable of beating even the Nazgul beasts in speed, so it seems like a plausible thought that this sentence is true. On the other hand, people argue that Sauron would have more easily seen them coming and that the eagles could not withstand direct confrontation with Sauron; therefore there are good reasons why this statement is, arguably, false.

---

[1] The authors would especially like to thank Aybüke Özgün, who has been an incredible help with the details of this paper. Thanks are also due to the audience at Logica 2017, Ilaria Canavotto, Thom van Gessel, and Anthia Solaki.

Tom Schoonen and Francesco Berto

This is one example of humans who reason about counterfactual circumstances *within fictional* situations. Reasoning about fiction is important because we investigate, through counterfactual speculation, how things might have gone otherwise in a fiction – e.g., when we want to expose a hole in the plot. More importantly, studying this kind of reasoning may give us clues on how mental simulation and counterfactual reasoning in general work. We will argue that, in counterfactual reasoning about fiction, three things happen: (1) one (sequentially) takes on board the explicit content of the fiction; (2) one imports background beliefs, which is represented by most plausible worlds after an update triggered by the explicit content; and (3) from these most plausible worlds, an agent reasons on alternative courses of action within the fiction.

We will proceed as follows: we start by briefly exposing the formal framework of Fontaine and Rahman (2014), which we take to be our starting model. We will then extend this to deal with the question of truth in fiction. In section 2, we will add a closeness ordering in order to deal with the counterfactual reasoning within fiction. After this, we will remark on some of the formal features of the resulting model. Finally, we will reflect on the work we have done and suggest some future research in the last section.

Before we get started, some preliminary remarks. In order to capture blatantly inconsistent fictions and, in particular, for the similarity ordering, we allow for impossible worlds. Impossible worlds can be such that are inconsistent (e.g., making both φ and $\neg\varphi$ true) or incomplete (e.g., making neither φ nor $\neg\varphi$ true). For any world w, let '$|w|$' denote the truth-set of that world in a model $\mathcal{M}$, i.e., $|w| =_{df} \{\varphi \mid \mathcal{M}, w \vDash \varphi\}$. We make two assumptions: (i) worlds always have a *non-empty* truth-set and (ii) for each set of sentences of the language, there is a(n impossible) world that makes those and only those sentences true. Conversely, for any sentence φ, let '$|\varphi|$' denote the truth-set of that sentence, i.e., $|\varphi| =_{df} \{w \mid \mathcal{M}, w \vDash \varphi\}$.[2]

1 Truth in fiction

We take as our starting point a model proposed by Fontaine and Rahman (2014) and suggest a few extensions. Fontaine and Rahman propose their

---

[2]This is sometimes called the proposition expressed by φ, however, we want to avoid using such a theoretically-loaded term. Also, note that it might be misleading to call situations that are incomplete 'impossible *worlds*', as the allocation 'world' seems to imply a form of maximality or completeness. However, we will follow the literature and refer to these incomplete situations also as impossible worlds (cf. Berto, 2013).

model as a formalisation of an *Artifactual Theory* of fictional objects. Such a theory claims that fictional objects are existent, created abstract objects. In order to capture all the insights of Artifactual Theorists, they formalise a variety of ontological dependencies. Due to space limitations, we will ignore part of the framework that is meant to deal with these dependencies and extend the model of Fontaine and Rahman (2014) with impossible worlds and a different accessibility-relation.

Note that the set of impossible worlds, I, and the set of possible worlds, P, are jointly exhaustive of the logical space and mutually exclusive. That is, $I \cup P = W$ and $I \cap P = \varnothing$.

DEFINITION 1. **Reasoning about Fiction Model**
Consider the following model, $\mathcal{M}$: $\langle P, I, \mathcal{A}, \mathcal{R}_\Phi, \{\preceq_a\}_{a \in \mathcal{A}}, V \rangle$:

▶ A set of *possible* worlds, P, and a set of *impossible* worlds, I,

▶ A set of agents, $\mathcal{A}$,

▶ An accessibility relation, $\mathcal{R}_\Phi$ on W, s.t. $\mathcal{R}_\Phi \subseteq W \times \wp(W)$,

▶ A set of agent-dependent plausibility orderings, $\preceq_a$, on all $S \subseteq W$,

▶ A standard valuation function, V.

Note that we have an accessibility relation for any intentional verb Φ, $\mathcal{R}_\Phi$, that we use when we discuss the worlds of a fiction. As there can be impossible fictions and fictional objects, this accessibility relation ranges of *all* worlds, also the impossible ones.[3] Most importantly, we have added an agent-dependent plausibility ordering to the model: $\preceq_a$. We take this to be a well-preorder. In the next section we will explain what this is, how it works, and how we will use it in our account of truth in fiction. Finally, the valuation function, V, takes a world and a sentence and returns '1' if the world is in the truth-set of the sentence and '0' otherwise. At possible worlds the truth-values are defined through the standard recursive definitions and at impossible worlds sentences are evaluated directly, without regard for compositionality.

---

[3]This is particular to a Modal Meinongian account of fictional objects: there is a world where the fictional object exist, not as fictional object, but as the thing it is described as being by the fiction (cf. Priest, 2005 and Berto, 2011). Note that, if one holds that fictional objects are *necessarily* non-existent, she could opt for letting $\mathcal{R}_\Phi$ range *only* over impossible worlds. Relatedly, in order to prevent concluding that actual objects are fictional objects due to accidental similarities, we could explicitly exclude the actual world from the range of $\mathcal{R}_\Phi$ (see Priest, 2005, p. 124).

Tom Schoonen and Francesco Berto

1.1 Truth in fiction with a plausibility ordering

Nichols and Stich (2000) use belief-revision in their cognitive theory of pretence; we take this as *prima facie* evidence that a realistic account of reasoning about fiction should include such a mechanism. So, our account of truth in fiction is based on Lewis' (1978) Analysis 2, extended with a belief-revision update on the beliefs of an agent.[4] The idea is something along the following lines: we take truth in fiction to be truth in the set of most plausible worlds for an agent after having updated her beliefs with the sentences of the fiction (or pretended to).[5] To this end, we use an agent-dependent plausibility-order on any $S \subseteq W$: $\preceq_a$ (see Def. 1).

Here, we will briefly explain how an agents arrives at a set of worlds she finds most plausible to be the worlds of the fiction, as these worlds will provide the starting point of our analysis of reasoning about fiction.

To see how we get to the set of most plausible worlds for an agent after having read a fiction, we need to go over some notational conventions. For instance, we let '$\bar{\varphi}$' denote a sequence of sentences, $\varphi_0 \varphi_1 \ldots \varphi_n$, and for any fiction, F, we let '$\mathcal{F}$' denotes the (finite) sequence of explicit sentences of that fiction. Secondly, let '$\mathcal{B}_a$' be an abbreviation of a specific instance of the accessibility relation: $\mathcal{R}^a_{\text{belief}}(@)$. That is, '$\mathcal{B}_a$' denotes the belief-set of agent a at the actual world. We will focus on the plausibility-ordering on the belief-set of an agent.

We follow Baltag and Smets (2006) in that the plausibility-order is a *well-preorder*. A *well*-preorder, is a preorder (i.e., reflexive and transitive) such that "every non-empty subset [of W] has minimal elements" (Baltag & Smets, 2006, p. 12, emphasis removed), where the minimal elements are defined as elements at least as plausible as all other elements of the set. We write '$w \preceq_a w'$' when an agent a considers world w at least as plausible as w'. Updating with a particular sentence, φ, will influence the plausibility-order. Let '$\preceq_a^\varphi$' denote the resulting plausibility-order after an update with φ that (i) preserves well-preorderedness; (ii) is such that $\forall w \in |\varphi|$ and $\forall w' \notin |\varphi|$, $w \preceq_a^\varphi w'$; and (iii) is such that the ordering within the φ-worlds and the non-φ-worlds remains the same.

There are two important things to note here. First of all, note that such an update makes all worlds that make φ true more plausible than all worlds

---

[4]We sloppily use 'update' for what, e.g., van Benthem (2007) calls *Lexicographic upgrade*.

[5]In our setting, an agent is engaging with fiction and the updates she performs are not actual, i.e., she engages in pretence. This could be incorporated formally by indexing the sentences and corresponding updates so that the agent can later 'unroll' the updates to recover her original plausibility-order. We will ignore this complication here.

that fail to make φ true; whether or not these worlds makes the negation of φ true is irrelevant. This is important as we work with impossible worlds. A stronger requirement would be that after such an update all the worlds that make φ true are more plausible than the worlds that fail to do so *and the worlds that make $\neg \varphi$ true*. For now, we will not impose this stronger requirement. Secondly, the ordering within the φ-worlds and the non-φ-worlds remains the same. To make things a bit more clear, consider a model with two propositional variables, p and q and a language with four sentences: $p, q, \neg p, \neg q$. Then, after an update with p, the set of most plausible worlds is as follows: $w_1 = \{p\}, w_2 = \{p, q\}, w_3 = \{p, \neg q\}, w_4 = \{p, \neg p\}, w_5 = \{p, q, \neg q\}, w_6 = \{p, q, \neg p\}, w_7 = \{p, \neg q, \neg p\}$, and $w_8 = \{p, \neg p, q, \neg q\}$.

We now have everything we need to define the set of most plausible worlds for an agent after having read a fiction, F. We denote this set with 'BEST$_{\preceq_a^F} \mathcal{B}_a$' ('BEST$_{\preceq_a^F}$' for short).

$$\text{BEST}_{\preceq_a^F} \mathcal{B}_a =_{df} \{w \in W \mid \forall w' \in \mathcal{B}_a [w \preceq_a^F w']\}$$

We take it that this set of most plausible worlds captures what the agents takes to be true in a particular fiction, F. (We will not give a clause for truth-in fiction; for a fully worked out account of this see Badura, 2016.) Note that this analysis deals both with steps (1) and (2) mentioned in the introduction. That is, given this analysis we are able to deal with the explicit content of the fiction (i.e., the sequential updates) as well as the relevant imported background information on the basis of this (i.e., the plausibility-order). In the next section, when dealing with counterfactual reasoning about fiction, we will concern ourselves with the last step, (3).[6]

2 Counterfactual reasoning about fiction

The main novelty of our paper will be the modelling of (3): counterfactual reasoning about fiction (henceforth: reasoning about fiction). So, for example, sentences such as 'If Sherlock Holmes had access to DNA-testing, then he would have solved the case of the Baskerville Hound much faster,'

---

[6]In what follows, we assume that there is some (objective) set of most plausible worlds after reading a fiction. These worlds should make true the explicit content of the fiction and some of the relevant background knowledge. However, this leaves open the option that one has her own account of how we get to the set of such worlds. That is, the analysis below is independent of our analysis of truth in fiction; all that the former assumes is that there is a set of worlds an agents takes to be candidates for the worlds of the fiction.

or 'If Frodo would have flown on an eagle, he would have reached Middle Earth sooner'. In order to model such reasoning, we draw inspiration from simulation models of counterfactual reasoning (cf. Nichols & Stich, 2000 and Williamson, 2007). In particular, we aim to formalise the insight that "[f]rom the initial premise along with her own current perceptions, her background knowledge, her memory of what has already happened in the episode, [...], *the pretender is able to draw inferences about what is going on in the pretense*" (Nichols & Stich, 2000, p. 119, emphasis added).

In order to use this insight we will proceed as follows: (i) we will use $\text{BEST}_{\preceq \mathcal{F}}$ to select the starting point from where the agent does the reasoning about the fiction; (ii) we will use an inference-based closesness-ordering of worlds (inspired by work of Mark Jago, 2009, 2014) to model the inferences the agent uses in her reasoning; (iii) and finally, we will give a semantics for a conditional, '$\boxed{F}\!\!\rightarrow$', that is to capture this reasoning about fiction.

So, first we need to find a suitable starting point from where the agent starts to reason about the fiction. We take it that when one evaluates a counterfactual about the fiction – which we will formulate as '$\varphi \boxed{F}\!\!\rightarrow \psi$' – she first updates $\text{BEST}_{\preceq \mathcal{F}}$, in the usual way, with the antecedent of the counterfactual. So, the set of most plausible worlds is now based on a sequential update of all the sentences of the fiction, i.e., $\mathcal{F}$, and finally with another update of the antecedent, φ. We will denote the result of this with '$\text{BEST}_{\preceq \mathcal{F}\varphi}$'. Clearly, the starting point of the agents reasoning should be related to this set. What will be important for (ii) is that the agent can 'reason towards' a conclusion, which we will model with a relation from one world to another, where the latter world makes more things true than the former. This way, the move from the one world to the other represents the fact that the agent inferred new information. Therefore, what we need as the starting point for the agent is a world that makes true the explicit content and the relevant background facts (i.e., is in $\text{BEST}_{\preceq \mathcal{F}\varphi}$), but that makes *no other things true*. That is, the world that makes those and only those things true. (It is here where the crucial need for incomplete worlds should become clear.)

That is, the 'smallest' world, in terms of the number of propositions it makes true, from the set of most plausible worlds after having updated with the fictional content, $\mathcal{F}$, and the antecedent. Let us call this the *minimal-set*. Formally:[7]

$$\text{MIN}^{\varphi}_{\mathcal{F}} =_{df} \{w \in W \mid \forall w' \in \text{BEST}_{\preceq \mathcal{F}\varphi}(|w| \not\supset |w'|)\}$$

---

[7]This is likely to be a singleton, however, it might also be an equivalence class of worlds. For what follows this does not matter significantly and we will often talk as if it is a singleton.

Reasoning About Fiction

This minimal world nicely captures the idea of Nichols and Stich that "[s]ome [...] constraints are imposed by the details of what has gone on earlier in the pretense along with the pretender's background knowledge. [However, t]his still *leaves many options open*" (2000, p. 127, emphasis added).

Secondly, we need a way to model (ii), i.e., the step-by-step reasoning towards the consequent. We will use an inference-based closeness-ordering to model this. That is, we will order worlds based on the number of inference steps 'it takes to get there' relative to the world of evaluation. Informally, given a particular information state (i.e., world), we can count the number of inference steps it would take to get from that information state to a new information state (world). The more inference steps needed to get to a world, the further away the world is. (Here, we follow work done on such orderings by Jago, 2009, 2014.)

So, we need to define the set of inferences that one is capable of making in reasoning from the minimal-world to other worlds. (Note that, as Jago (2009) already points out, having the ability to apply certain inference rules is not the same as having one's beliefs closed under those inference rules.) Let $\mathfrak{R}$ be the set of rules that one can easily make when reasoning about fiction. (For now, we will not go into the details of what rules that might be.) This set of inference rules allows us to model reasoning about fiction with the following formal tools.

First of all, given the set of rules, we can define a set of ordered-pairs of worlds such that the first world of the pairs makes true the premises of an argument and the latter world additionally makes true the conclusion that we can draw with the particular inference rule. So, let $[\![\mathfrak{R}]\!]$ be the set of ordered-pairs of worlds, then $\langle w, w' \rangle$ is in that set iff there is a rule in $\mathfrak{R}$,

$$\frac{\varphi_1, \ldots \varphi_n}{\psi} \; r \in \mathfrak{R}$$

such that $\{\varphi_1, \ldots \varphi_n\} \in |w|$, $\psi \notin |w|$, and $|w'| = \{\psi\} \cup |w|$. For what follows, we need two weak assumptions with respect to the inference rules that possibly are in $\mathfrak{R}$:

Assumption 1: For all sets of inference rules (used here), it is not the case that there is an inference rule $r \in \mathfrak{R}$ such that $\varphi \wedge \neg \varphi \vdash_r \psi$.

Assumption 2: For all sets of inference rules (used here), there is an inference rule $r \in \mathfrak{R}$ such that $\varphi \wedge \neg \varphi \vdash_r \varphi$ and $\varphi \wedge \neg \varphi \vdash_r \neg \varphi$.

We take it that these are *prima facie* reasonable assumptions. The first denies the following reasoning: if there is an object that is green all over and red all over that it follows that Sherlock Holmes never solved a crime. The second asserts that from the fact that there is something that is both round and square one can reason to the information that there is something that is round.

Given the minimal-world and the set of inference rules, we now finally have everything in place to define a function that will provide us with the closeness-order. As noted above, we assume that there is a particular antecedent of the counterfactual and this influences what is in the minimal-set, therefore the function is in a sense sentence-specific (so for every sentence of our language there will be such a function):

> Let f_φ be a partial function from worlds, W, to the natural numbers, $W \to \mathbb{N}$. Then, $f_\varphi(w) = n$ if and only if there is a sequence of worlds, $w_0 w_1 \ldots w_n$, such that (i) $w_0 = w$, (ii) $w_n \in \text{MIN}_{\mathcal{F}}^\varphi$, and (iii) for all $i \leq n$, $\langle w_i, w_{i-1} \rangle \in [\![\mathfrak{R}]\!]$, and there is no such sequence $w_0 w_1 \ldots w_m$ for $m < n$.

This function does exactly what we intuitively want it to do. Given a world, this function determines the smallest number of inference steps from the minimal-world to that given world.

With this closeness-order, we can now finally turn to (iii), giving a semantics for the conditional that aims to capture reasoning about fiction: $\varphi \boxminus\!\!\to \psi$. Again, we aim to capture the insights from cognitive science and epistemology concerning counterfactual reasoning. Nichols and Stich describe the process of reasoning about pretence as follows: "[e]arly on in a typical episode of pretense, [...], one or more initial pretense premises are placed in the [...] workspace. [...] What happens next is that the cognitive system *starts to fill* the [scenario] with an *increasingly detailed* description of what the world would be like if the initiating representation were true" (2000, p. 122, emphases added). Williamson uses this description to provide an epistemology of counterfactuals and argues that "one asserts the counterfactual conditional if and only if [such] development eventually leads one to add the consequent" (2007, p. 153).

We follow Williamson in our semantics, namely, that the conditional is true if, when reasoning from the antecedent, one reaches the consequent before she reaches its negation. Formally:

DEFINITION 2. **Semantic-clause f-counterfactual**
Consider a fiction, F, and a world, $w \in W$, then $\mathcal{M}, w \models \varphi \boxminus\!\!\rightarrow \psi$ iff $\exists w'(f_\varphi(w')$ is defined, $\mathcal{M}, w' \models \psi$, and $\forall w''[\mathcal{M}, w'' \models \neg\psi \Rightarrow f_\varphi(w') \leq f_\varphi(w'')])$

Let us call this conditional, '$\boxminus\!\!\rightarrow$', the *f-counterfactual* for future reference. Note that the universal quantifier used here makes it that if there is no world that makes the negation of the consequent true, then the f-counterfactual is vacuously true. However, the existential quantifier makes it such that if there is no world where the consequent is true, the counterfactual will be false.

3 Non-monotonicity, reflexivity, and modus ponens

Before we conclude with some more philosophical reflections on our analysis, let us briefly remark three formal features of the conditional.

In general with impossible worlds semantics or logic, validity is defined as truth-preservation at all *possible* worlds. Most classical validities are remained that way. However, if we want to look at the behaviour of our conditional, non-normal worlds are crucial, as we saw. So, when we talk about what 'holds' for our conditional, we mean at all worlds, even impossible ones.

Non-Monotonicity: First of all, one of the main features of counterfactual reasoning is that it is a form of *non-monotonic* reasoning, that is, additional information in the antecedent might make false an otherwise true counterfactual. A classic example that shows this is the following: 'if I had struck this match, it would have lit' is true, yet 'if I had struck this match and it had been wet, it would have lit' is false. It is quite straightforward to see that our conditional, $\boxminus\!\!\rightarrow$, captures such non-monotonicity. Informally, this is explained by the fact that we use a sequential update for the set $\text{BEST}_{\leq^F}$. This means that our ordering is sensitive to the most recent sentence with which it is updated. Thus, it is not at all necessary that the minimal-world MIN_F^φ is the same as the minimal-world $\text{MIN}_F^{\varphi\chi}$. As these two worlds may be different, the consequences that can be drawn from these worlds need not be identical. Hence, the conditional allows for non-monotonic reasoning.

Reflexivity: Another feature that holds for our conditional is that of *reflexivity*, i.e., $\varphi \boxminus\!\!\rightarrow \varphi$. This is also rather straightforward. Consider an arbitrary φ and we can show that $\varphi \boxminus\!\!\rightarrow \varphi$. By definition of '$\text{MIN}^\varphi$' and the update procedure, we know that all worlds in MIN^φ are such that they make φ true and that there is at least one such world. Then, by definition of '$\boxdot\!\!\rightarrow$', we know what $\varphi \boxdot\!\!\rightarrow \varphi$ is true, for after an update with the antecedent, all worlds in MIN$^\varphi$ already make the consequent true. Thus, the conditional is reflexive.

Modus Ponens: A feature that many theorists agree on to hold for counterfactuals is *modus ponens*, that is, the following inference should hold:

$$\varphi, \varphi \boxdot\!\!\rightarrow \psi \vDash \psi$$

As we pointed out above, when talking about modus ponens, we will not talk of validity. Even more so, it is important to note that modus ponens generally does not hold at impossible worlds for *any* conditional. Consider any language and model with incomplete worlds, then we can always construct a world where the conditional and the antecedent are true, yet that is 'ignorant' on the status of the consequent. Similarly, for any language and model with inconsistent worlds, we can always construct a world where the conditional and the antecedent hold, but where the negation of the consequent is true.

So, we will here focus on a particular kind of, what we will call, 'restricted modus ponens'. That is, we will prove that there is a characteristic of this conditional that one can interpret as a version of modus ponens.[8] In order to do so, we will first prove a lemma about $\mathfrak{R}$ and then we will define a function from worlds to worlds, that given a world, delivers the closure of that world under the inference rules in $\mathfrak{R}$. First the lemma:

LEMMA 3.1. If an ordered pair of worlds, $\langle w, w' \rangle$, is in the set $[\![\mathfrak{R}]\!]$, then all propositions made true by the former world will also be made true by the latter, i.e., $\forall \varphi \in |w|[\varphi \in |w'|]$.

Proof. Let w and w' be such that $\langle w, w' \rangle \in [\![\mathfrak{R}]\!]$. It follows by definition of $\mathfrak{R}$ that the truth-set of w' is the same as the truth-set of w, with the addition of one sentence, ψ, that is not in w. So, $|w| \subset |w'|$. Hence, all truths from w are preserved in w'. □

This lemma will be used later on when we prove our form of modus ponens, but first, we need to define a function on worlds, that gives us the closed version of worlds under the rules of $\mathfrak{R}$:

---

[8] Thanks to Mark Jago for his suggestions on modus ponens and the needed functions and thanks to Aybüke Özgün for many corrections on the formal details.

Let $f^{\mathfrak{R}}$ be a function from worlds, W, to worlds, $W \to W$. Then, $f^{\mathfrak{R}}(w) = w'$ if and only if there is a sequence $w_0 w_1 \ldots w_n$ such that (i) $w_0 = w$; (ii) for all $i < n$ it is the case that $\langle w_i, w_{i+1} \rangle \in [\![\mathfrak{R}]\!]$; (iii) w_n is such that $\forall \varphi (|w_n| \vdash_{r \in \mathfrak{R}} \varphi \Rightarrow \varphi \in |w_n|)$; and (iv) $w' = w_n$.

This function takes a world and gives the 'closure-world' of that world under the rules of inference in $\mathfrak{R}$. Note that $f^{\mathfrak{R}}(\cdot)$ need not give us consistent worlds as the rules of inference in $\mathfrak{R}$ are non-monotonic, nor need they be complete, as some facts might be left out from the start and/or be non-inferable.

Before we can start our proof, note that in the definition of f_φ, the ordering is 'from the world with the larger truth-set *to the world with the smaller* truth-set'. On the other hand, in the definition of $f^{\mathfrak{R}}$, the ordering is the other way around, that is, 'from the world with the smaller truth-set *to the world with the larger* truth-set'. Luckily, it is easy to see that, *given* a sequence, we can re-label that sequence: instead of naming the starting world w_0 and the last world w_n, we call the starting world w_n and the last world w_0. That is, we 'flip' the ordering, but only of the labels we use for the worlds. The 'objective' order of the worlds is unaffected by this. As the labels are mere names for the worlds, this is fine as long as we then reverse the definition of which worlds follow from which worlds. So:

LEMMA 3.2. If there is a sequence of worlds, $w_0 w_1 \ldots w_n$ such that (i) $w_0 = w$, (ii) $w_n \in \text{MIN}^\varphi_{\mathcal{F}}$, and (iii) for all $i \leq n, \langle w_i, w_{i-1} \rangle \in [\![\mathfrak{R}]\!]$, then there is a sequence of worlds, $w_0 w_1 \ldots w_n$ such that (i) $w_n = w$, (ii) $w_0 \in \text{MIN}^\varphi_{\mathcal{F}}$, and (iii) for all $i \leq n, \langle w_i, w_{i+1} \rangle \in [\![\mathfrak{R}]\!]$.

Proof. Follows immediately from re-labelling. □

We will give a small toy-example to make this intuitive. Consider worlds u, v, w such that $[\![\mathfrak{R}]\!] = \{\langle u, v \rangle, \langle v, w \rangle\}$. If we now reason from u to w, there is a sequence, $w_0 w_1 w_2$ such that $w_0 = u$, $w_2 = w$, and for each i in the sequence $\langle w_i, w_{i+1} \rangle \in [\![\mathfrak{R}]\!]$. However, we can reason similarly from w to u. In that case, there is a sequence, $w_0 w_1 w_2$ such that $w_0 = w$, $w_2 = u$, and for each i in the sequence $\langle w_i, w_{i-1} \rangle \in [\![\mathfrak{R}]\!]$.

With all this in place, it is quite straightforward to show what kind of 'restricted modus ponens' holds for our conditional.

THEOREM 3.1. For all models, $\mathcal{M}$, all fictions, F, and all worlds, w, if $w \in \text{BEST}_{\preceq F}$, then, if $w \vDash \varphi, \varphi \boxdot\!\!\!\rightarrow \psi$, $f^{\mathfrak{R}}(w) \vDash \psi$.

Proof. Take an arbitrary world, w, and assume that $w \vDash \varphi$ and that $w \vDash \varphi \boxdot\!\!\!\rightarrow \psi$. By definition of the plausibility-ordering, the fact that w makes true φ, and the $w \in \text{BEST}_{\preceq F}$ it follows that $w \in \text{BEST}_{\preceq F\varphi}$. From this, and the definition of $\text{MIN}^{\varphi}_{\mathcal{F}}$, it follows that the truth-set of all worlds, w', that are in $\text{MIN}^{\varphi}_{\mathcal{F}}$, are a subset of the truth-set of w. That is, $\forall w' \in \text{MIN}^{\varphi}_{\mathcal{F}}(|w'| \subseteq |w|)$.

Now, from the fact that $w \vDash \varphi \boxdot\!\!\!\rightarrow \psi$ and the semantics of '$\boxdot\!\!\!\rightarrow$', it follows that there is a world, let's name it u, such that $u \vDash \psi$ and such that there is a sequence, $w_0 w_1 \ldots w_n$ such that (i) $w_0 = u$, (ii) $w_n \in \text{MIN}^{\varphi}_{\mathcal{F}}$, and (iii) for all $i \leq n, \langle w_i, w_{i-1} \rangle \in [\![\mathfrak{R}]\!]$. By Lemma 3.2, we can 'flip' this ordering for ease of exposition such that there is a sequence of worlds $w_0 w_1 \ldots w_n$ such that (i) $w_n = u$, (ii) $w_0 \in \text{MIN}^{\varphi}_{\mathcal{F}}$, and (iii) for all $i \leq n, \langle w_i, w_{i+1} \rangle \in [\![\mathfrak{R}]\!]$. What this shows is that u is 'reachable' from $\text{MIN}^{\varphi}_{\mathcal{F}}$ with the rules of $\mathfrak{R}$ and, as we saw, $u \vDash \psi$.

By the definition of $f^{\mathfrak{R}}$, $f^{\mathfrak{R}}(w)$ gives us a sequence of worlds $w_0 w_1 \ldots w_n$, such that (i) $w_0 = w$, (ii) for all $i \leq n$ it is the case that $\langle w_i, w_{i+1} \rangle \in [\![\mathfrak{R}]\!]$; (iii) w_n is such that $\forall \varphi(|w_n| \vdash_{r \in \mathfrak{R}} \varphi \Rightarrow \varphi \in |w_n|)$; and (iv) $f^{\mathfrak{R}}(w) = w_n$.

Now, remember that the truth-set of all the worlds in $\text{MIN}^{\varphi}_{\mathcal{F}}$ are a subset of the truth-set of w. So, if we can $\mathfrak{R}$-derive u from $\text{MIN}^{\varphi}_{\mathcal{F}}$, it follows that we can also $\mathfrak{R}$-derive u from w. And, given the 'closedness' of $f^{\mathfrak{R}}(w)$ under the rules in $\mathfrak{R}$, we know that there is a $i \leq n$, such that in the relevant sequence, $w_i = u$.

Given Lemma 3.1, we know that $|u| \subseteq |f^{\mathfrak{R}}(w)|$. This, combined with the fact that $u \vDash \psi$, gives us that $f^{\mathfrak{R}}(w) \vDash \psi$. □

Remember that this is *not* the classical modus ponens as the consequent is not made true by the same world as the antecedent and the conditional. However, if the world in question is a *possible* world, this would be a version of classical modus ponens; because possible worlds are deductively closed, so for any possible world, $w = f^{\mathfrak{R}}(w)$.[9]

We do believe that this version of modus ponens is intuitive enough though. For consider what one does when she reasons about fiction, she

---

[9]However, note that this requires getting into the debate concerning the possibility of possible worlds being the most plausible worlds as the 'fictional world'. We will not engage with this here, we only want to note this technicality.

does indeed start to reason from the world she took to be the most plausible 'fiction world'. However, given the new input of the antecedent, this world might no longer be the most plausible one (also given the interaction of the antecedent with the relevant background beliefs).

Also, briefly note why we need the assumptions mentioned above. As we already mentioned, it might very well be that $f^{\mathfrak{R}}(w)$ results in an inconsistent world such that $f^{\mathfrak{R}}(w) \vDash \psi$ and $f^{\mathfrak{R}}(w) \vDash \neg\psi$. However, given **Assumption 2**, it still follows that $f^{\mathfrak{R}}(w) \vDash \psi$. Conversely, if there is an inconsistent world in the sequence, we do not want that from that moment onwards anything can be inferred. **Assumption 1** secures this.

4 Conclusion: counterfactual reasoning and $\mathfrak{R}$

The above model for counterfactual reasoning about fiction is far from full-fledged qua technical details. However, we believe that it provides a valuable proof of concept. That is, we think that this type of analysis of counterfactuals matches an intuitive account of the epistemology and cognitive processes involved in counterfactual reasoning (cf. Nichols & Stich, 2000 and Williamson, 2007, Ch. 5). We hope to explore such accounts of reasoning more fully in the future, especially in relation to counterfactual reasoning in general, but also other forms of conditional reasoning (see for example, Solaki & Berto, 2017, who apply a similar analysis to dual-processing theories of reasoning).

We will conclude this paper by making some remarks on the set of inference rules, $\mathfrak{R}$, which does much of the heavy lifting on our account.

An important question that remains unanswered is what (kinds of) rules of inferences are in $\mathfrak{R}$. This, we take it, is a very difficult question, one that we cannot hope to settle here. We want to briefly remark two things with regards to it – one methodological note and one suggesting an avenue of relevant research – hoping to relieve some of the pressure of this question.

First of all, note that the lack of specific content of $\mathfrak{R}$ does not present a knock-down argument against the account presented here. Surely, more needs to be said about the specifics of $\mathfrak{R}$ and the details of the current proposal probably need to be adapted in accordance. However, one might wonder whether this is an issue for logicians to address. In order to have a cognitively realistic model, determining what inferences people make, especially in the context of fictive counterfactual reasoning, should be left to cognitive scientists. The logician should have a framework ready in which she can incorporate the parameters provided by the cognitive scientists (see also Jago, 2014 on this).

These remarks only relieve the pressure of the question a bit and we can in fact make some suggestions for future research. Let us mention two. First of all, it seems clear that people, in general, do not reason by the standards of classical logic, let alone when reasoning about fictions. For example, Priest (2005) notes that it is likely that paraconsistent logic "is the default logic for reasoning about a fictional situation" (p. 122). Similarly, relevant logics, non-monotonic logics, conditional logics, etc., might all prove to involve certain inferences that agents make in the context of fiction. Secondly, it seems clear that agents do not make inferences on the basis of their information *unrestrictedly* (cf. Jago, 2014; Solaki, 2017; Solaki & Berto, 2017). That is, there are *cognitive costs* to making such inferences – if only the fact that human agents are finite beings. Solaki (2017) provides a wide range of models (most based on dynamic epistemic logics with impossible worlds) that keep track of such cognitive costs (see also Solaki & Berto, 2017). Interestingly, the models she uses also involve rational agents making step-by-step inferences from their information. We take it that research into (i) non-classical logics and (ii) the cognitive costs of such inference steps will provide valuable insights that are needed to further work out the details concerning $\mathfrak{R}$.

We take the above not to be vices of the analysis we provide here, but as exciting new venues of research that will hopefully lead to a new accounts of counterfactuals, counterpossibles, and conditional reasoning in general that closely matches findings from areas of cognitive theories and the epistemology of such reasoning.

References

Badura, C. (2016). *Truth in Fiction via Non-Standard Belief Revision* (Unpublished master's thesis). ILLC, University of Amsterdam.

Baltag, A., & Smets, S. (2006). Dynamic Belief Revision over Multi-Agent Plausibility Models. In *Proceedings of the 7th Conference on Logic and the Foundations of Game and Decision (LOFT2006)* (p. 11-24).

van Benthem, J. (2007). Dynamic logic for belief revision. *Journal of Applied Non-Classical Logics*, *17*(2), 129-155.

Berto, F. (2011). Modal Meinongianism and fiction: the best of three worlds. *Philosophical Studies*, *152*(3), 313-334.

Berto, F. (2013). Impossible Worlds. In E. N. Zalta (Ed.), *The Stanford Encyclopedia of Philosophy*. Stanford, CA.: CSLI Publications.

Fontaine, M., & Rahman, S. (2014). Towards a semantics for the artifactual theory of fiction and beyond. *Synthese, 191*(3), 499-516.

Jago, M. (2009). Logical information and epistemic space. *Synthese, 167*, 327-341.

Jago, M. (2014). *The Impossible*. Oxford: Oxford University Press.

Lewis, D. K. (1978). Truth in fiction. *American Philosophical Quarterly, 15*(1), 37-46.

Nichols, S., & Stich, S. (2000). A cognitive theory of pretense. *Cognition, 74*, 115-147.

Priest, G. (2005). *Towards Non-Being; the logic and metaphysics of intentionality*. Oxford: Oxford University Press.

Solaki, A. (2017). *Steps out of Logical Omniscience* (Unpublished master's thesis). ILLC, University of Amsterdam.

Solaki, A., & Berto, F. (2017). The Logic of Fast and Slow Thinking. *Unpublished manuscript*.

Williamson, T. (2007). *The Philosophy of Philosophy*. Oxford: Blackwell Publishing.

Tom Schoonen
Institute for Logic, Language, and Computation
The Netherlands
E-mail: T.Schoonen@uva.nl

Francesco Berto
Institute for Logic, Language, and Computation
The Netherlands
E-mail: F.Berto@uva.nl

A Logical Perspective on Social Group Creation

SONJA SMETS AND FERNANDO R. VELÁZQUEZ-QUESADA

Abstract: This extended abstract introduces a research line on a logical approach to the different mechanisms via which social groups are created. We start with a basic setting for describing the agents' features and social network, after which we explore different alternatives for social network creation as well as the particular perspectives that arise in each of the cases.

Keywords: Social influence, Social network, Social network creation, Epistemic logic, Dynamic epistemic logic

1 Introduction

It is well-known that our social environment affects both our behaviour and our opinions about the world. As different studies in social psychology have shown (e.g., Crisp, 2015; Fiske, Gilbert, & Lindzey, 2010), our traits, intentions, behaviour, beliefs and preferences are influenced by the (actual, imagined or implied) presence of others. *Social influence* can take many forms, as socialisation (inheriting and disseminating norms, customs, values and ideologies), conformity (changing attitudes, beliefs and behaviours to match those of the majority), compliance (changing favourably in response to explicit or implicit requests made by others), reactance (adopting a view contrary to what the person is being pressured to accept) and obedience (changing in response to a direct command from an authority figure).[1] These phenomena, also studied by theoretical sciences (e.g., network and graph theories: Diestel, 2017; Estrada, 2011; agent-based modelling: Naldi, Pareschi, & Toscani, 2010; Shoham & Leyton-Brown, 2009; social network analysis: Easley & Kleinberg, 2010; McCulloh, Armstrong, & Johnson, 2013), have become even more important now, given the rise of the internet and virtual societies.

---

[1] See, e.g., Cialdini and Griskevicius (2010); Nowak, Vallacher, and Miller (2013); see Cialdini (2008) for a general audience presentation.

Sonja Smets and Fernando R. Velázquez-Quesada

The usual approach to the formal study of social influence is to use quantitative modelling, a tradition that can be traced back to French (1956), with DeGroot (1974) and Lehrer and Wagner (1981) being some of the most significant proposals. Still, these are not the only formal tools that can be used to analyse social dynamics. In particular, *epistemic logic* (*EL*; Fagin, Halpern, Moses, & Vardi, 1995; Hintikka, 1962) and *dynamic epistemic logic* (*DEL*; Baltag, Moss, and Solecki 1998; van Benthem 2011; van Ditmarsch, van der Hoek, and Kooi 2008) have proven to be useful tools when studying the way agents interact with each other. While the first has given rise to formal studies of multi-agent epistemic concepts as common knowledge (Barwise, 1988; Halpern & Moses, 1984) and distributed knowledge (Halpern & Moses, 1985), the second has given rise to the study of different forms of agents' communication, including communication networks and protocols (Apt, Grossi, & van der Hoek, 2015; Baltag, 2001; van Ditmarsch, van Eijck, Pardo, Ramezanian, & Schwarzentruber, 2017; Sietsma & van Eijck, 2011; Wang, Sietsma, & van Eijck, 2010).

Recently, these logical tools have been employed to study the subtler forms of interaction that fall under the umbrella of social influence. One of the earlier papers is Seligman, Liu, and Girard (2011), which explores social information flow within social networks. Several other papers did follow, as Zhen and Seligman (2011) with its analysis of peer pressure, Baltag, Christoff, Hansen, and Smets (2013) with its study of informational cascades, Liu, Seligman, and Girard (2014) with its study of belief change in social networks, Christoff, Hansen, and Proietti (2016) with its study of reflective social influence, Baltag, Christoff, Rendsvig, and Smets (2016) with its research on diffusion and prediction update, and Velázquez-Quesada (2017) with its exploration on priority-based peer influence, among others. These approaches tend to use qualitative tools instead of quantitative ones, and instead of looking at 'complex' data describing influence behaviour, they rely on relatively 'simple' models, using formal languages to describe general 'complex' patterns about the agents' interaction.

Still, while the importance of social groups is acknowledged in logical studies, the way these groups are created has received less attention.[2] This is important, as the interaction between the social-group-influencing-individual and individuals-creating-social-groups phenomena plays a crucial role in the way a group and its member will behave.

---

[2] Some exceptions are Pedersen and Slavkovik (2017); Solaki, Terzopoulou, and Zhao (2016).

A Logical Perspective on Social Group Creation

The goal of this extended abstract is to introduce a research line on the different mechanisms through which social networks are built. The proposal starts (Section 2) with our basic setting, representing an agent in terms of her *binary* 'yes/no' choices for the different *features* she may have. Our setting further follows a *similarity* approach, according to which the more similar two agents are, the more likely it is for them to end up in the same social network. There are other alternatives that one could choose for these two basic choices. For example, while each agent's feature may be described in more detail via a *number* within the $[0, 1]$ interval, whether they will be part of the same social group might be decided not by how similar they are but rather by how well they *complement* each other. Still, our chosen basic setting already allows for the study of several interesting possibilities and extensions.

The first of these alternatives is the precise way this *similarity* approach is carried out (Section 3). This paper's idea is to use a notion of *distance*, stating 'how similar' two given agents are. But this still gives us room to play, as the distance may depend not only on the 'issues' (features) that are 'being discussed' at each given moment, but also on how important each one of them is for the involved agents. The second is the precise way in which the distance between agents is used to define the social group (Section 4). These notes discuss both a *threshold* and a *group-size* approach, with each one of them yielding networks with different topological properties. A third crucial point concerns any further ingredients that should be taken into account when creating a social network (Section 5); these might include social requirements ("my potential future friends are current friends *of my friends*") as well as epistemic ones ("my potential future friends should be people who *I know* have similarities with me"). Finally, it is interesting to look not only at the result of a single step, but rather at the result of an infinite number of them. In this line, Section 6 discusses some questions about the social network's shape 'in the limit', examining also those that arise when the network-construction and social influence dynamics are put together.

2 Social network models

This paper's basic model (cf. Baltag et al., 2016) is a relational 'Kripke' model, with its domain interpreted as the set of agents, its accessibility relation representing the social connection, and its atomic valuation describing the features each agent has. Let A denote a finite non-empty set of agents, and P a finite set of features each agent might or might not have.

Definition 1 (Social network model) *A social network model (SNM) is a tuple $M = \langle \mathsf{A}, S, V \rangle$ where $S \subseteq \mathsf{A} \times \mathsf{A}$ is the social relation (Sab indicates that agent a is socially connected to agent b) and $V : \mathsf{A} \to \wp(\mathsf{P})$ is a feature function ($p \in V(a)$ indicates that agent a has feature p).*

These structures can be described by a propositional *language $\mathcal{L}$*, built up from special atoms describing the agents' features and their social relationship. More precisely, formulas in $\mathcal{L}$ are given by

$$\varphi, \psi ::= p_a \mid \mathsf{S}_{ab} \mid \neg \varphi \mid \varphi \wedge \psi$$

with $p \in \mathsf{P}$ and $a, b \in \mathsf{A}$. We read p_a as "agent a has feature p" and S_{ab} as "agent a is socially connected to b". *Other Boolean operators (with $\veebar$ the exclusive disjunction) are defined as usual. Given a SNM $M = \langle \mathsf{A}, S, V \rangle$, the semantic interpretation of formulas in $\mathcal{L}$ is given by*

$$M \Vdash p_a \;\; \mathit{iff}_{\mathit{def}} \;\; p \in V(a), \qquad M \Vdash \neg \varphi \;\; \mathit{iff}_{\mathit{def}} \;\; M \not\Vdash \varphi,$$
$$M \Vdash \mathsf{S}_{ab} \;\; \mathit{iff}_{\mathit{def}} \;\; Sab, \qquad M \Vdash \varphi \wedge \psi \;\; \mathit{iff}_{\mathit{def}} \;\; M \Vdash \varphi \text{ and } M \Vdash \psi.$$

A formula $\varphi \in \mathcal{L}$ is valid ($\Vdash \varphi$) when $M \Vdash \varphi$ holds for all models M.

As opposed to other approaches (e.g., Baltag et al., 2016; Christoff et al., 2016; Liu et al., 2014), in a SNM the relation S is not required to satisfy any specific property; in particular, it needs to be neither irreflexive nor symmetric. This allows for the representation of more realistic scenarios, as some agents might think of themselves as 'friends' and, moreover, friendship preferences (unfortunately) do not need to be mutual.

Multiple networks. As we will show, a single social network already provides interesting aspects to be discussed. Yet, a simple generalisation is to consider rather a collection of them. In order to give these networks a stronger meaning, one can move to a setting in which each feature (e.g., "music") provides not a choice between *"yes"* and *"no"*, but rather a choice among a finite range of values (e.g., *"rock"*, *"classic"* and *"jazz"*). Then, one can define a social network for each feature $p \in \mathsf{P}$, and agents can be grouped according to the value they assign to each such p. After all, someone who chooses football as her favourite sport and Lady Gaga as her favourite musician is bound to have different social environments in each one of these contexts. Multiple networks will be important when considering the long-term behaviour of the network building process, as discussed in Section 6.

A Logical Perspective on Social Group Creation

3 Distance between agents

Distance. As mentioned above, the chosen approach for social network construction is that of *similarity*: the more similar two agents are, the more likely it is for them to become socially related. A formal definition of this idea relies on a notion of 'distance' between agents, counting the number of features in which they differ.[3]

Definition 2 (Distance) *Let* $M = \langle A, S, V \rangle$ *be a SNM; define the set of features distinguishing agents* $a, b \in A$ *in* M *as* $\text{MSMTCH}^M(a, b) := \big(V(a) \setminus V(b)\big) \cup \big(V(b) \setminus V(a)\big)$. *The function* $\text{DIST}^M : A \times A \to \mathbb{N}$, *indicating the distance in* M *between two agents, is given by*

$$\text{DIST}^M(a, b) := |\text{MSMTCH}^M(a, b)|$$

Note how DIST is indeed a mathematical distance: for any agents $a, b \in$ A,*(i)* the distance from a to b is non-negative, *(ii)* the distance from a to b is equal to that from b to a, and *(iii)* the distance from an agent to herself is 0. Moreover, DIST is a *semi-metric*, as 'going' directly from a to c is at least as 'fast' as 'going' via another agent (*subadditivity*). Still, DIST is not a *metric*, as it does not satisfy *identity of indiscernibles*: $\text{DIST}^M(a, b) = 0$ does not imply $a = b$, as two different agents may have exactly the same features.

Relative distance. A proper notion of distance is enough for proposing similarity approaches to social network creation. However, its definition is not without alternatives. As an example, one might understand these social-network-construction scenarios as 'public conversations' where all agents 'discuss' their features. Then, as the 'conversation' continues, agents will form subgroups of people sharing common characteristics. From this perspective, it is clear that not all features can be 'discussed' at once: just some of them will be relevant at each stage of the interaction. This is not a novel idea: Solaki et al. (2016) uses a game theoretic setting to define the agreement and disagreement of agents on a specific feature (or issue), and this yields a way for them to update the social relation of agents with respect to one specific feature at a time. Below is our alternative.

Definition 3 (Q-Distance) *Let* $M = \langle A, S, V \rangle$ *be a SNM; let* $Q \subseteq P$ *be a set of features. The Q-distance between* a *and* b *in* M *(that is, the distance between* a *and* b *in* M *relative to features in Q) is given by*

$$\text{DIST}^M_Q(a, b) := |\text{MSMTCH}^M(a, b) \cap Q|$$

---
[3]This is the *Hamming* distance, but others (e.g., the Jaccard distance) are also possible.

Sonja Smets and Fernando R. Velázquez-Quesada

This refined notion of distance (Smets & Velázquez-Quesada, 2017a) allows us to describe more realistic scenarios, such as the step-by-step interaction in real dialogues (when personal features are slowly revealed as the conversation goes on), or cases in which agents control when one of their features becomes visible to other agents (e.g. when agents choose to expose some 'private' information only in specific circumstances).

Subjective distance. On the other hand, for these two *objective* notions of distance, what matters is the number of differences, and not what these differences are. But this is hardly the case in real life, some features may be more relevant than others. Maybe more importantly, the 'degree of importance' tends to differ from agent to agent: vegetarianism might be, for some but not for all, a more important attribute than political affiliation. This suggests not only a notion of distance that relies on a given *priority* among features, but also one in which this priority differs from agent to agent.

Definition 4 (Extended social network model) An extended SNM (XSNM) is a tuple $M = \langle \mathsf{A}, S, \{\preccurlyeq_a\}_{a \in \mathsf{A}}, V \rangle$ where $M = \langle \mathsf{A}, S, V \rangle$ is a SNM and each $\preccurlyeq_a \subseteq (\mathsf{P} \times \mathsf{P})$ is a total preorder on features ($p \preccurlyeq_a q$ indicates that, for agent a, feature p is at most as important as feature q)[4]

In an XSNM, a distance between agents can be defined as the *weighted sum* of the features in which the agents differ, with this weight possibly differing from agent to agent (Smets & Velázquez-Quesada, 2017b).

Definition 5 (Subjective distance) Let $M = \langle \mathsf{A}, S, \{\preccurlyeq_a\}_{a \in \mathsf{A}}, V \rangle$ be an XSNM. For defining a subjective distance between agents, note first how each $\preccurlyeq_a$ induces an ordered sequence of subsets of P ($\mathrm{P}_0^a, \mathrm{P}_1^a, \ldots$), each one containing features that are, from a's perspective, equally important (P_0^a contains $\preccurlyeq_a$'s most important features). Moreover, P is finite, so only a finite number of such sets (say, $\sharp_a$) will be non-empty. Then, by using a function assigning weight to features in each P_k^a (say, $\mathrm{WEIGHT}_a(k) := \sharp_a - k$), the distance between agents b_1 and b_2 in M according to agent a is given by

$$\mathrm{DIST}_a^M(b_1, b_2) := \sum_{k=0}^{\sharp_a - 1} \left(|\mathrm{MSMTCH}^M(b_1, b_2) \cap \mathrm{P}_k^a| \cdot \mathrm{WEIGHT}_a(k) \right)$$ [5]

---
[4]Thus, every two features are comparable (by $\preccurlyeq_a$'s totality), and there are maximum ones (by P's finiteness).

[5]Note how, with such definition of WEIGHT, while the weight of the most important features (those in P_0^a) is $\sharp_a$, that of the least important (those in $\mathrm{P}_{\sharp_a - 1}^a$) is 1.

A Logical Perspective on Social Group Creation

Complex features. The definitions of distance provided so far are measured in terms of the elements of P. But agents could also be interested in *combinations* of these basic traits. For example, in the latter case of agents having priorities over features, an agent might assign a certain weight u_v to vegetarianism and a certain weight u_{pa} to political activism, and yet assign a weight $u > u_v + u_{pa}$ to their simultaneous occurrence. Then, she might also assign *negative* points to agents lacking certain attributes. The distance between agents need not be measured only in terms of basic features: it can also be measured in terms of their *Boolean* combinations.

But in this case it is natural to think of other kinds of features. Some of them may be social. For example, an agent might be more comfortable within small groups, therefore ranking higher those agents with few friends. Others might want to 'avoid' a certain agent c, and thus rank higher those agents that are not socially connected with her (those b for which Sbc is not the case). Even more, the aversion for c might be really strong, thus preferring not those who are not friends of c, but rather those that are not socially connected to her (those b for which S^+bc is not the case, with S^+ the transitive closure of S). The first two situations can be described by our language, as there are $\mathcal{L}$-formulas expressing not only that a given agent a has exactly f friends[6], but also that she is not related to c. The third situation can be described by adding a modality S^+_{ab} whose semantic interpretation relies on S^+.

Other features may be *epistemic* (I prefer agents that *know* their own – basic– features): those can be handled within an epistemic social network setting (Definition 9). There are also dynamic features, as an agent's aversion for c may be so strong that she would prefer no agents that are not socially connected to c, but rather agents that *will never be* socially connected to her. Such requirements (and many others, as having 'stable' social environments) can be handled only after introducing the actual methods via which social networks will change.

4 Different group-creation strategies

Once a notion of distance between agents has been fixed, the next step is to use it to build a social network. This proposal follows the *DEL* approach, representing dynamic actions as operations that change the model. Here are some of the possibilities

---

[6] $\bigvee_{\{B \subseteq A : |B|=f\}} \left(\bigwedge_{b \in B} S_{ab} \wedge \bigwedge_{c \in A \setminus B} \neg S_{ac} \right)$

Threshold approach. A natural option is to let two agents become friends when they are 'similar' enough. If a *threshold* $\theta \in \mathbb{N}$ is given, with $\theta < |\mathsf{P}|$ (recall: P is finite), then we can define a *threshold-similarity* update operation allowing agents to establish connections with those who differ in at most θ features (Smets & Velázquez-Quesada, 2017).

Definition 6 (Threshold update) *Let* $M = \langle \mathsf{A}, S, V \rangle$ *be a SNM; take* $\theta \in \mathbb{N}$. *The* threshold-similarity update *of M generates the SNM* $M_{\odot_\theta} = \langle \mathsf{A}, S_{\odot_\theta}, V \rangle$, *with its social relation given by*

$$S_{\odot_\theta} := \{(a,b) \in \mathsf{A} \times \mathsf{A} \,:\, \text{DIST}^M(a,b) \leq \theta\} \qquad 7$$

Thus, each agent defines a circle of ratio θ with herself at the center, and her social contacts will be those agents falling inside it. For the language, the dynamic modality $[\odot_\theta]$ describes the changes this operation brings about:

$$M \Vdash [\odot_\theta]\varphi \quad \text{iff}_{def} \quad M_{\odot_\theta} \Vdash \varphi.$$

When the threshold update is used, the resulting social network will be both reflexive and symmetric:

$$\Vdash [\odot_\theta] S_{aa}, \qquad \Vdash [\odot_\theta](S_{ab} \to S_{ba}).$$

However, transitivity and Euclideanity cannot be guaranteed:

The initial SNM appears on the left, and the one that results from the threshold-similarity update with $\theta = 1$ appears on the right. Transitivity fails because what distinguishes a from b ($\{q\}$) is *only part of* what distinguishes a from c ($\{p,q\}$); Euclidianity fails because what distinguishes b from a ($\{q\}$) is *different from* what distinguishes b from c ($\{p\}$).

Group-size approach. There are alternatives to the threshold strategy. An interesting idea, arising from the cognitive science literature, is to take into account the size of the agent's 'social space'. In real life, agents may be willing to keep expanding their social network (even including people who are

---

[7]For an XSNM, an appropriate definition would use $\text{DIST}^M_a(a,b)$.

A Logical Perspective on Social Group Creation

very different from them) as long as there is still 'enough space' in their social environment.[8] This is famously known as the *Dunbar's number*: a suggested cognitive limit to the number of people with whom one can maintain stable social relationships (see, e.g., Dunbar, 1992), and it suggests an alternative for social network creation (Smets & Velázquez-Quesada, 2017a).

Definition 7 (Group-size update) *Let* $M = \langle A, S, V \rangle$ *be a SNM; take* $\lambda \in \mathbb{N}$. *The function* DIST *induces an ordered sequence of subsets of* A $(A^a_{-1}, A^a_0, A^a_1, \ldots)$, *with all but the first containing agents equally distant from agent* a $(A^a_0$ *contains the closest ones), and the first an empty set used for technical reasons. Moreover, denote by* $\flat_a(\lambda)$ *the 'last' layer of contacts* $a \in A$ *can add to her network without going above* λ. *The* group-size-similarity update *of* M *generates the SNM* $M_{\bowtie_\lambda} = \langle A, S_{\bowtie_\lambda}, V \rangle$, *with its social relation given by*

$$S_{\bowtie_\lambda} := \{(a,b) \in A \times A \,:\, b \in \bigcup_{k=-1}^{\flat_a(\lambda)} A^a_k\} \qquad [9]$$

Thus, each agent starts adding agents to her social network, starting with the closest ones, and finishing once her 'social space' reaches its limit λ. *For describing the changes this operation brings about,*

$$M \Vdash [\bowtie_\lambda] \varphi \quad \textit{iff}_{def} \quad M_{\bowtie_\lambda} \Vdash \varphi.$$

Note how, since each A^a_i might have more than one element, each agent could reach a point where adding agents in the next set would take her above the limit; thus, in some cases agents will have some 'memory slots' empty. The additional set A^a_{-1} makes the definition work in cases in which A^a_0 contains already too many agents; in such situations, after a group-size update, the agent will be 'friendless'.

Interestingly, this approach produces social relations with different properties. First, an agent will consider herself as part of her new social network if and only if the amount of agents that are feature-wise identical to her is at most the limit λ. Maybe more interesting, the following diagrams show how neither symmetry nor transitivity need to hold (take $\lambda = 2$).[10]

---

[8] Think, for example, how, when close friends are not around, we establish conversations with relatively 'distant' acquaintances.

[9] For an XSNM, an appropriate definition would use $\text{DIST}^M_a(a,b)$.

[10] Numbers over edges indicate distance. Edges in black are actual pairs in the social network relation, and dotted grey edges are shown only for distance information.

279

Sonja Smets and Fernando R. Velázquez-Quesada

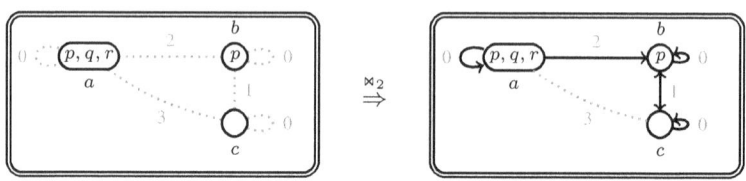

In general, symmetry fails because a high occurrence of similar agents will create fully connected clusters, leaving dissimilar agents with asymmetric edges. Such failure of symmetry also leads to the failure of transitivity: by their own space limitations, these 'stranded' agents will be connected to only some of the agents in the cluster.

This does not mean that interesting relational properties can never be achieved. For example, the social relation will be symmetric when the agents are 'similarly dissimilar', that is, when their differences are uniformly distributed. In such cases, the update will yield ring-like structures with symmetric edges, as shown in diagram below ($\lambda = 3$).

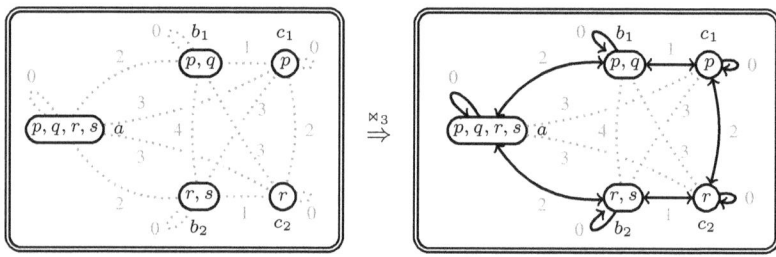

These examples highlight a crucial difference between the threshold and the group-size approach. Under the threshold method, what matters for deciding whether Sab will be the case (besides θ) is only the distance between a and b. However, under the group-size method, what matters for deciding whether Sab will be the case (besides λ) is the distance between a and *all agents*. For example, a and b might be extremely similar (say, $\text{DIST}^M(a,b) = 1$), and still b will not be in a's social network if the number of agents feature-wise identical to a is high.[11] More precisely, the group-size approach is *context-sensitive*: the social connections agent a will have do not depend on 'how fit' each candidate is individually, but rather on how fit each candidate is *compared with the rest*. Group-size is not about similarity; it is about *relative* similarity.

---

[11] Even more: a and b might be feature-wise identical, and yet they will not be socially connected if the number of agents feature-wise identical to them is larger than the group-size.

A Logical Perspective on Social Group Creation

5 Further requirements

The two explored social network creation alternatives have a common point: they rely only on the *similarity* of the agents (i.e., the distance between them), leaving every other aspect out. However, other requirements could be called for. Some of them might be *social*: after all, a large part of the new people we meet are friends of our friends. Some others might be *epistemic*: after all, two 'identical souls' will relate to each other only after they know about their similarities. The discussion below (cf. Smets & Velázquez-Quesada, 2017) uses the threshold approach, but others (as the group-size alternative) are also possible.

Social requirements. In several social scenarios, it takes a common acquaintance (i.e., a *middle man*) to introduce new friends.

Definition 8 (Middleman threshold update) *Let $M = \langle \mathsf{A}, S, V \rangle$ be a SNM; take $\theta \in \mathbb{N}$. The* middleman threshold-similarity update *of M generates the SNM $M_{\widehat{\odot}_\theta} = \langle \mathsf{A}, S_{\widehat{\odot}_\theta}, V \rangle$, with its social relation given by*

$$S_{\widehat{\odot}_\theta} := \{(a,b) \in \mathsf{A} \times \mathsf{A} : \mathrm{DIST}^M(a,b) \leq \theta \text{ and } \exists\, c \in \mathsf{A} \text{ s.t. } Sac \And Scb\}$$

For describing the changes this operation brings about,

$$M \Vdash [\widehat{\odot}_\theta]\varphi \quad \textit{iff}_{def} \quad M_{\widehat{\odot}_\theta} \Vdash \varphi.$$

In the new network, agent a will include agent b as a social contact if and only if they are similar enough *and* there is an agent c who belongs to a's social network and whose social network includes b. Social requirements for the middleman might vary. In some cases, a *symmetric* social relation between her and both a and b might be required; in others, an agent who has both a and b in her social network might be enough.

The middleman requirement changes the properties of the resulting network. For example, as opposed to the basic threshold case, reflexivity and symmetry might fail: the prescribed middleman might not exist. Moreover, just as there might be no middlemen for establishing new relations, there might be no middlemen for *preserving* old ones; in this variation, the social network is not monotone. Of course, monotonicity can be enforced by defining the new social relation 'accumulatively' ($S_{\widehat{\odot}_\theta} := S \cup \cdots$). Yet, this is not always appropriate: in real scenarios, social connections can be created, but unfortunately (and, in some cases, fortunately) they can be lost too. An advantage of not enforcing monotonicity is that it is possible to identify those situations that lead to it naturally. Here, a *reflexive* S guarantees that old social contacts will be preserved (modulo the agents' distance).

$$\Vdash (S_{ab} \wedge \mathrm{Dist}_{ab}^{\leq \theta} \wedge S_{aa}) \to [\widehat{\odot}_\theta] \, S_{ab} \qquad {}^{12}$$

As the well-known saying goes, *"if you want to be loved, you should love yourself first"*. In our setting, this becomes *"one way to preserve your close friends is to think of yourself as worthwhile to be friend of"*.

Epistemic requirements. Sometimes, what matters is not only how similar two agents are, but also whether they know about this similarity. In order to incorporate such epistemic requirements, the model should be extended.

Definition 9 (Epistemic social network model) *An epistemic social network model (**ESNM**) is a standard multi-agent relational (possible worlds) model (Hintikka, 1962) in which each possible world w stands for a **SNM** $\langle \mathsf{A}, S_w, V_w \rangle$ and the epistemic relations $\sim_i$ are equivalence relations.*[13]

For describing these structures, the modal language $\mathcal{L}^{\mathrm{K}}$ includes not only $\mathcal{L}$'s special atoms, but also epistemic modalities K_a for each agent a:

$$\varphi, \psi ::= p_a \mid S_{ab} \mid \neg\varphi \mid \varphi \wedge \psi \mid \mathrm{K}_a \, \varphi$$

As usual, a formula $\mathrm{K}_a \, \varphi$ is read as "*agent a knows φ*". The semantic interpretation is standard for Boolean operators and the epistemic modalities, with atoms p_a and S_{ab} interpreted relative to evaluation point:

$$(M, w) \Vdash p_a \; \mathit{iff}_{def} \; p \in V_w(a), \qquad (M, w) \Vdash S_{ab} \; \mathit{iff}_{def} \; S_w ab.$$

For epistemic requirements in social network creation, here is an option.

Definition 10 (Knowledge-based threshold Update) *Let $M = \langle W, \mathsf{A}, \sim, S, V \rangle$ be an **ESNM**; take $\theta \in \mathbb{N}$. The knowledge-based threshold similarity update operation generates the **ESNM** $M_{\odot_\theta^{\mathrm{K}}} = \langle W, \mathsf{A}, \sim, S_{\odot_\theta^{\mathrm{K}}}, V \rangle$, with*

$$(S_{\odot_\theta^{\mathrm{K}}})_w := \{(a, b) \in \mathsf{A} \times \mathsf{A} : \forall u \sim_a w \, , \, \mathrm{dist}_u^M(a, b) \leq \theta\}$$

Syntactically,

$$(M, w) \Vdash [\odot_\theta^{\mathrm{K}}] \varphi \quad \mathit{iff}_{def} \quad (M_{\odot_\theta^{\mathrm{K}}}, w) \Vdash \varphi.$$

---

[12] With $\mathrm{Dist}_{ab}^{\leq \theta} := \bigvee_{\{Q \subseteq P \, : \, |Q|=t\}} \left(\bigwedge_{p \in Q} (p_a \veebar p_b) \wedge \bigwedge_{p \in P \setminus Q} (p_a \leftrightarrow p_b) \right)$ an $\mathcal{L}$-formula stating that a and b differ in at most θ features (Smets & Velázquez-Quesada, 2017).

[13] Constrains can be imposed: one can ask for the agents to know themselves ($w \sim_a u$ implies $V_w(a) = V_u(a)$) or to know who are her contacts ($w \sim_a u$ implies $S_w[a] = S_u[a]$).

A Logical Perspective on Social Group Creation

This operation extends the basic threshold approach (Definition 6) with an epistemic requirement: in world w agent a will add b to her social network if and only if in this world a *knows* that her distance to b fits the threshold. The resulting networks will be reflexive, even in those cases in which the agent 'does not know herself.[14] On the other hand, social relations might not be symmetric, as the agents' knowledge might not have such property: at w agent a may know that she and b are similar enough (so $(S_{\odot_\theta^K})_w ab$ will hold), but b may not know this (and thus $(S_{\odot_\theta^K})_w ba$ will fail).

Bringing epistemic requirements to the table also brings the *de dicto* and *de re* distinction. The proposed operation uses a *de dicto* approach: after the operation, a includes b in her network when a knows that b is similar enough, even if she does not know which are their shared features.

$$\Vdash [\odot_\theta^K] S_{ab} \leftrightarrow K_a \text{Dist}_{ab}^{\leq \theta}$$

On the other hand, a *de re* approach would ask not for a to know that she and b are similar enough, but rather for her to point out a large enough set of features on which she and b coincide.

$(S'_{\odot_\theta^K})_w := \{(a,b) \in A \times A : \exists Q \subseteq P \text{ s.t.}$
 (i) $|Q| \geq |P| - \theta$ and *(ii)* $\forall u \sim_a w$, $V_u(a) \cap Q = V_u(b) \cap Q \}$

$$\Vdash [\odot_\theta^K] S_{ab} \leftrightarrow \bigvee_{t=|P|-\theta}^{|P|} \bigvee_{\{Q \subseteq P : |Q|=t\}} K_a \bigwedge_{p \in Q}(p_a \leftrightarrow p_b)$$

This *de re* version is a better fit for agents that give different features a different priority (Definition 4), as they would want to know *which* the differences and similarities are before making a 'friendship' decision.

Added value of epistemic aspects. Epistemic requirements raise further alternatives. Their combination with the relative and subjective distances of Section 3 suggests situations where strategic behaviour plays an important role. If agent a knows that she and b differ in the feature b considers the most important, she (a) might want to keep this topic out of the conversation, at least until it has been commonly established (i.e., it is common knowledge between a and b) that they nevertheless have some similarities.

Epistemic and social requirements may also be combined. For example, new social links may require not only similarity but also a middleman and knowledge of the agents about both. Social requirements in an epistemic setting also present a *de dicto* vs *de re* choice: either there is someone that a

---

[14] $\Vdash [\odot_\theta^K] S_{aa}$: in any possible world, the distance between any agent and herself is 0.

knows can link her with b, or else a only knows there is someone who can link her with b. But then, the epistemic burden can be shifted to the middleman: a and b will be socially related if and only if there is a middleman who is not only socially related to both agents, but also knows they are similar enough (what online dating sites and dating apps do).

Then, an obvious next step is to study knowledge changing operations (e.g., public and private announcements) within ESNM models, focussing not only on the way the agents' knowledge about each other's features and social connections change, but also on its interplay with the described knowledge-based social network changing operations. Additionally, there are interesting situations in which agents can learn new facts about each other's features and social relations from both the knowledge they have about the group-formation rules and the way the network has changed.

6 Long term

Note how the setting given in both the basic threshold and the basic group-size approaches is 'idempotent': new networks depend only on the agents's basic features and thus, as they do not change, after the network has been updated, further updates will not make any difference. This changes not only when social requirements are added (the middlemen of Section 5) but also when more complex features are considered (size of social environment, social connection –or lack of– with a given agent). In such settings, one can wonder whether a infinite sequence of update operations might reach a 'stable' point, and if so, under which conditions.

Finally, an important topic is the interplay between the feature-based social-network-changing operations of this research project, and the social-network-based feature-changing operations (social influence) mentioned in the introduction. Both ideas deserve to be studied in tandem: the dynamics of one can affect the dynamics of the other, and our logical setting might be able to capture interesting properties concerning their interplay. For example, the social phenomenon of *polarisation* can be understood as a process through which agents sharing similar views form a social group, and then the group's 'echo chamber' leads them to have a more extreme position. This in turn might lead to the alienation of some, which leaves only the most radical ones in the group, at which point the cycle repeats itself. Here, being part of different social networks might prove to be helpful, as the different 'gravity force' of the different networks might help to keep the agent 'balanced'.

References

Apt, K. R., Grossi, D., & van der Hoek, W. (2015). Epistemic protocols for distributed gossiping. In R. Ramanujam (Ed.), *Proceedings Fifteenth Conference on Theoretical Aspects of Rationality and Knowledge, TARK 2015, Carnegie Mellon University, Pittsburgh, USA, June 4-6, 2015.* (Vol. 215, pp. 51–66).

Baltag, A. (2001). Logics for insecure communication. In J. van Benthem (Ed.), *Proceedings of the 8th Conference on Theoretical Aspects of Rationality and Knowledge (TARK-2001), Certosa di Pontignano, University of Siena, Italy, July 8-10, 2001* (pp. 111–122). Morgan Kaufmann.

Baltag, A., Christoff, Z., Hansen, J. U., & Smets, S. (2013). Logical models of informational cascades. In J. van Benthem & F. Liu (Eds.), *Logic Across the University: Foundations and Applications. Proceedings of the Tsinghua Logic Conference, Beijing, 2013* (Vol. 47, pp. 405–432). London: College Publications.

Baltag, A., Christoff, Z., Rendsvig, R. K., & Smets, S. (2016). Dynamic epistemic logics of diffusion and prediction in social networks (extended abstract). In *Proceedings of LOFT 2016.*

Baltag, A., Moss, L. S., & Solecki, S. (1998). The logic of public announcements, common knowledge, and private suspicions. In I. Gilboa (Ed.), *Proceedings of TARK-98* (pp. 43–56).

Barwise, J. (1988). Three views of common knowledge. In M. Y. Vardi (Ed.), *TARK II* (pp. 365–379). Morgan Kaufmann.

van Benthem, J. (2011). *Logical Dynamics of Information and Interaction.* Cambridge University Press.

Christoff, Z., Hansen, J. U., & Proietti, C. (2016). Reflecting on social influence in networks. *Journal of Logic, Language and Information,* 25(3-4), 299–333.

Cialdini, R. B. (2008). *Influence: Science and practice* (5th ed.). Boston, USA: Allyn & Bacon.

Cialdini, R. B., & Griskevicius, V. (2010). Social influence. In R. F. Baumeister & E. J. Finkel (Eds.), *Advanced Social Psychology: The State of the Science* (pp. 385–417). New York, USA: Oxford University Press.

Crisp, R. J. (2015). *Social Psychology: A Very Short Introduction.* Oxford, UK: Oxford University Press.

DeGroot, M. H. (1974). Reaching a consensus. *Journal of the American Statistical Association,* 69(345), 118–121.

Diestel, R. (2017). *Graph Theory* (5th ed., Vol. 173). Berlin, Heidelberg: Springer-Verlag.

van Ditmarsch, H., van der Hoek, W., & Kooi, B. (2008). *Dynamic Epistemic Logic* (Vol. 337). Dordrecht, The Netherlands: Springer.

van Ditmarsch, H., van Eijck, J., Pardo, P., Ramezanian, R., & Schwarzentruber, F. (2017). Epistemic protocols for dynamic gossip. *Journal of Applied Logic*, *20*, 1–31.

Dunbar, R. I. M. (1992). Neocortex size as a constraint on group size in primates. *Journal of Human Evolution*, *22*(6), 469–493.

Easley, D., & Kleinberg, J. (2010). *Networks, Crowds and Markets: Reasoning about a Highly Connected World*. New York: Cambridge University Press.

Estrada, E. (2011). *The Structure of Complex Networks: Theory and Applications*. New York: Oxford University Press.

Fagin, R., Halpern, J. Y., Moses, Y., & Vardi, M. Y. (1995). *Reasoning about knowledge*. Cambridge, Mass.: The MIT Press.

Fiske, S. T., Gilbert, D. T., & Lindzey, G. (Eds.). (2010). *Handbook of Social Psychology* (5th ed.). New Jersey, USA: Wiley. (Volumes 1 & 2)

French, J. R. P. (1956). A formal theory of social power. *Psychological Review*, *63*(3), 181–194.

Halpern, J. Y., & Moses, Y. (1984). Knowledge and common knowledge in a distributed environments. In *PODC '84: Proceedings of the third annual ACM symposium on Principles of distributed computing* (pp. 50–61). New York, N.Y., U.S.A.: ACM.

Halpern, J. Y., & Moses, Y. (1985). A guide to the modal logics of knowledge and belief: Preliminary draft. In A. K. Joshi (Ed.), *Proceedings of the 9th International Joint Conference on Artificial Intelligence. Los Angeles, CA, USA, August 1985* (pp. 480–490). Morgan Kaufmann.

Hintikka, J. (1962). *Knowledge and Belief: An Introduction to the Logic of the Two Notions*. Ithaca, N.Y.: Cornell University Press.

Lehrer, K., & Wagner, C. (1981). *Rational Consensus in Science and Society. A Philosophical and Mathematical Study* (Vol. 24). Dordrecht, Holland: Dordrecht Reidel Publishing Company.

Liu, F., Seligman, J., & Girard, P. (2014). Logical dynamics of belief change in the community. *Synthese*, *191*(11), 2403–2431.

McCulloh, I., Armstrong, H., & Johnson, A. (2013). *Social Network Analysis with Applications*. New Jersey, USA: John Wiley & Sons, Inc.

Naldi, G., Pareschi, L., & Toscani, G. (Eds.). (2010). *Mathematical Modeling of Collective Behavior in Socio-Economic and Life Sciences.* Boston: Birkhäuser.

Nowak, A., Vallacher, R. R., & Miller, M. E. (2013). Social influence and group dynamics. In H. Tennen & J. Suls (Eds.), *Handbook of Social Psychology* (2nd ed., Vol. 5: Personality and Social Psychology, pp. 383–417). New Jersey, USA: Wiley.

Pedersen, T., & Slavkovik, M. (2017). Formal models of conflicting social influence. In B. An, A. L. C. Bazzan, J. Leite, S. Villata, & L. W. N. van der Torre (Eds.), *PRIMA 2017: Principles and Practice of Multi-Agent Systems - 20th International Conference, Nice, France, October 30 - November 3, 2017, Proceedings* (Vol. 10621, pp. 349–365). Springer.

Seligman, J., Liu, F., & Girard, P. (2011). Logic in the community. In M. Banerjee & A. Seth (Eds.), *Logic and Its Applications - 4th Indian Conference, ICLA 2011, Delhi, India, January 5-11, 2011. Proceedings* (Vol. 6521, pp. 178–188). Springer.

Shoham, Y., & Leyton-Brown, K. (2009). *Multiagent Systems - Algorithmic, Game-Theoretic, and Logical Foundations.* Cambridge University Press.

Sietsma, F., & van Eijck, J. (2011). Message passing in a dynamic epistemic logic setting. In K. R. Apt (Ed.), *Proceedings of the 13th Conference on Theoretical Aspects of Rationality and Knowledge (TARK-2011), Groningen, The Netherlands, July 12-14, 2011* (pp. 212 220). ACM.

Smets, S., & Velázquez-Quesada, F. R. (2017). How to make friends: A logical approach to social group creation. In A. Baltag, J. Seligman, & T. Yamada (Eds.), *Logic, Rationality, and Interaction - 6th International Workshop, LORI 2017, Sapporo, Japan, September 11-14, 2017, Proceedings* (Vol. 10455, pp. 377–390). Springer.

Smets, S., & Velázquez-Quesada, F. R. (2017a). *The Creation and Change of Social Networks: a logical study based on group size.* (To appear in A. Madeira and M. Benevides, editors, Proceedings of Dynamic Logic: new trends and applications DALí 2017)

Smets, S., & Velázquez-Quesada, F. R. (2017b). *A logical study of agents' distances in social network creation.* (Proceedings of LAMAS 2017)

Solaki, A., Terzopoulou, Z., & Zhao, B. (2016). Logic of closeness revision: Challenging relations in social networks. In M. Köllner & R. Ziai (Eds.), *Proceedings of ESSLLI 2016 student session* (pp. 123–134).

Velázquez-Quesada, F. R. (2017). Reliability-based preference dynamics:

Lexicographic upgrade. *Journal of Logic and Computation*.

Wang, Y., Sietsma, F., & van Eijck, J. (2010). Logic of information flow on communication channels. In D. Grossi, L. Kurzen, & F. R. Velázquez-Quesada (Eds.), *Logic and Interactive RAtionality. Seminar's yearbook 2009* (pp. 226–245). Amsterdam, The Netherlands: Institute for Logic, Language and Computation, Universiteit van Amsterdam.

Zhen, L., & Seligman, J. (2011). The dynamics of peer pressure. In H. van Ditmarsch, J. Lang, & S. Ju (Eds.), *Logic, Rationality, and Interaction - Third International Workshop, LORI 2011, Guangzhou, China, October 10-13, 2011. Proceedings* (Vol. 6953, pp. 390–391). Springer.

Sonja Smets
Universiteit van Amsterdam
Institute for Logic, Language and Computation
The Netherlands
E-mail: `S.J.L.Smets@uva.nl`

Fernando R. Velázquez-Quesada
Universiteit van Amsterdam
Institute for Logic, Language and Computation
The Netherlands
E-mail: `F.R.VelazquezQuesada@uva.nl`

www.ingramcontent.com/pod-product-compliance
Lightning Source LLC
Chambersburg PA
CBHW050130170426
43197CB00011B/1784